Barracoon

ALSO BY ZORA NEALE HURSTON

Jonah's Gourd Vine

Their Eyes Were Watching God

Moses, Man of the Mountain

Seraph on the Suwanee

Mules and Men

Tell My Horse

Dust Tracks on a Road

Barracoon

The Story of the
Last "Black Cargo"

Zora Neale Hurston
Edited by Deborah G. Plant

HARPER LUXE

An Imprint of HarperCollins*Publishers*

Grateful acknowledgment is made for permission to reprint the following material:

Door of No Return. Courtesy of Deborah G. Plant and Gloria Jean Plant Gilbert

Kossula. Courtesy of McGill Studio Collection, The Doy Leale McCall Rare Book and Manuscript Library, University of South Alabama

HarperCollins books may be purchased for educational, business, or sales promotional use. For information please e-mail the Special Markets Department at SPsales@harpercollins.com.

FIRST HARPERLUXE EDITION

ISBN: 978-0-06-286436-9

HarperLuxe™ is a trademark of HarperCollins Publishers.

Library of Congress Cataloging-in-Publication Data is available upon request.

18 19 20 21 22   ID/LSC   10 9 8 7 6 5 4 3

But the inescapable fact that stuck in my craw, was: my people had *sold* me and the white people had bought me. . . . It impressed upon me the universal nature of greed and glory.

—ZORA NEALE HURSTON, *Dust Tracks on a Road*

Barracoon: The Spanish word *barracoon* translates as "barracks" and is derived from *barraca*, which means "hut." The term "barracoon" describes the structures used to detain Africans who would be sold and exported to Europe or the Americas. These structures, sometimes also referred to as factories, stockades, corrals, and holding pens, were built near the coast. They could be as insubstantial as a "slave shed" or as fortified as a "slave house" or "slave castle," wherein Africans were forced into the cells of dungeons beneath the upper quarters of European administrators. Africans held in these structures had been kidnapped, captured in local wars and raids, or were trekked in from the hinterlands or interior regions across the continent. Many died in the barracoons as a consequence of their physical condition upon arrival at the coast or the length of time it took for the arrival of a ship. Some died while waiting for a ship to fill, which could take three to six months. This phase of the traffic was called the "coasting" period. During the years of suppression of the traffic, captives could be confined for several months.

Contents

Afterword
and Additional Materials
Edited by Deborah G. Plant

Foreword

Those Who Love Us
Never Leave Us Alone
with Our Grief

Reading *Barracoon:*
The Story of the Last "Black Cargo"

Those who love us never leave us alone with our grief. At the moment they show us our wound, they reveal they have the medicine. *Barracoon: The Story of the Last "Black Cargo"* is a perfect example of this.

I'm not sure there was ever a harder read than this, for those of us duty bound to carry the ancestors, to work for them, as we engage in daily life in different parts of the world where they were brought in chains.

And where they, as slaves to cruel, or curious, or indifferent, white persons (with few exceptions) existed in precarious suspension disconnected from their real life, and where we also have had to struggle to protect our humanity, to experience joy of life, in spite of everything evil we have witnessed or to which we have been subjected.

Reading *Barracoon*, one understands immediately the problem many black people, years ago, especially black intellectuals and political leaders, had with it. It resolutely records the atrocities African peoples inflicted on each other, long before shackled Africans, traumatized, ill, disoriented, starved, arrived on ships as "black cargo" in the hellish West. Who could face this vision of the violently cruel behavior of the "brethren" and the "sistren" who first captured our ancestors? Who would want to know, via a blow-by-blow account, how African chiefs deliberately set out to capture Africans from neighboring tribes, to provoke wars of conquest in order to capture for the slave trade people—men, women, children—who belonged to Africa? And to do this in so hideous a fashion that reading about it two hundred years later brings waves of horror and distress. This is, make no mistake, a harrowing read.

We are being shown the wound.

However, Zora Hurston's genius has once again produced a Maestrapiece. What is a Maestrapiece? It is the
feminine perspective or part of the structure, whether
in stone or fancy, without which the entire edifice is
a lie. And we have suffered so much from this one:
that Africans were only victims of the slave trade, not
participants. Poor Zora. An anthropologist, no less!
A daughter of Eatonville, Florida, where truth, what
was real, what actually happened to somebody, mat-
tered. And so, she sits with Cudjo Lewis. She shares
peaches and watermelon. (Imagine how many generations of black people would never admit to eating
watermelon!) She gets the grisly story from one of the
last people able to tell it. How black people came to
America, how we were treated by black and white.
How black Americans, enslaved themselves, ridiculed
the Africans; making their lives so much harder. How
the whites simply treated their "slaves" like pieces of
machinery. But machinery that could be whipped if it
didn't produce enough. Fast enough. Machinery that
could be mutilated, raped, killed, if the desire arose.
Machinery that could be cheated, cheerfully, without a
trace of guilt.

And then, the story of Cudjo Lewis's life after Eman-
cipation. His happiness with "freedom," helping to cre-

ate a community, a church, building his own house. His tender love for his wife, Seely, and their children. The horrible deaths that follow. We see a man so lonely for Africa, so lonely for his family, we are struck with the realization that he is naming something we ourselves work hard to avoid: how lonely we are too in this still foreign land: lonely for our true culture, our people, our singular connection to a specific understanding of the Universe. And that what we long for, as in Cudjo Lewis's case, is gone forever. But we see something else: the nobility of a soul that has suffered to the point almost of erasure, and still it struggles to be whole, present, giving. Growing in love, deepening in understanding. Cudjo's wisdom becomes so apparent, toward the end of his life, that neighbors ask him to speak to them in parables. Which he does. Offering peace.

Here is the medicine:

That though the heart is breaking, happiness can exist in a moment, also. And because the moment in which we live is all the time there really is, we can keep going. It may be true, and often is, that every person we hold dear is taken from us. Still. From moment to moment, we watch our beans and our watermelons grow. We plant. We hoe. We harvest. We share with neighbors. If a young anthropologist appears with two hams and gives us one, we look forward to enjoying it.

Life, inexhaustible, goes on. And we do too. Carrying our wounds and our medicines as we go.

Ours is an amazing, a spectacular, journey in the Americas. It is so remarkable one can only be thankful for it, bizarre as that may sound. Perhaps our planet is for learning to appreciate the extraordinary wonder of life that surrounds even our suffering, and to say Yes, if through the thickest of tears.

Alice Walker
March 2018

Introduction

On December 14, 1927, Zora Neale Hurston took the 3:40 p.m. train from Penn Station, New York, to Mobile, to conduct a series of interviews with the last known surviving African of the last American slaver—the *Clotilda*. His name was Kossola, but he was called Cudjo Lewis. He was held as a slave for five and a half years in Plateau-Magazine Point, Alabama, from 1860 until Union soldiers told him he was free. Kossola lived out the rest of his life in Africatown (Plateau).[1] Hurston's trip south was a continuation of the field trip expedition she had initiated the previous year.

Oluale Kossola had survived capture at the hands of Dahomian warriors, the barracoons at Whydah (Ouidah), and the Middle Passage. He had been enslaved, he had lived through the Civil War and the largely un-Reconstructed South, and he had endured the rule of Jim Crow. He had experienced the dawn of a new millennium that included World War I and the Great Depression. Within the magnitude of world events swirled the momentous events of Kossola's own personal world.

Zora Neale Hurston, as a cultural anthropologist, ethnographer, and folklorist, was eager to inquire into his experiences. "I want to know who you are," she approached Kossola, "and how you came to be a slave; and to what part of Africa do you belong, and how you fared as a slave, and how you have managed as a free man?" Kossola absorbed her every question, then raised a tearful countenance. "Thankee Jesus! Somebody come ast about Cudjo! I want tellee somebody who I is, so maybe dey go in de Afficky soil some day and callee my name and somebody say, 'Yeah, I know Kossula.'"[2]

Over a period of three months, Hurston visited with Kossola. She brought Georgia peaches, Virginia hams, late-summer watermelons, and Bee Brand insect powder. The offerings were as much a currency to facilitate their blossoming friendship as a means to encourage Kossola's reminiscences. Much of his life was "a sequence of separations."[3] Sweet things can be palliative. Kossola trusted Hurston to tell his story and transmit it to the world. Others had interviewed Kossola and had written pieces that focused on him or more generally on the community of survivors at Africatown. But only Zora Neale Hurston conducted extensive interviews that would yield a comprehensive, book-length account of Kossola's life. She would alternately title the work "Barracoon: The Story of the Last 'Black Cargo'" and "The Life of Kossula." As with the other interviews, Kossola hoped the story he entrusted to Hurston would reach his people, for whom he was still lonely. The disconnection he experienced was a source of continuous distress.

ORIGINS

Kossola was born circa 1841, in the town of Bantè, the home to the Isha subgroup of the Yoruba people of West Africa. He was the second child of Fondlolu, who was the second of his father's three wives. His mother named him Kossola, meaning "I do not lose my fruits anymore" or "my children do not die any more."[4] His mother would have four more children after Kossola, and he would have twelve additional siblings from his extended family. Fondlolu's name identified her as one who had been initiated as an Orìṣà devotee. His father was called Oluale.[5] Though his father was not of royal heritage as *Olu*, which means "king" or "chief," would imply, Kossola's grandfather was an officer of the king of their town and had land and livestock.

By age fourteen, Kossola had trained as a soldier, which entailed mastering the skills of hunting, camping, and tracking, and acquiring expertise in shooting arrows and throwing spears. This training prepared him for induction into the secret male society called *oro*. This society was responsible for the dispensation of justice and the security of the town. The Isha Yoruba of Bantè lived in an agricultural society and were a peaceful people. Thus, the training of young men in the art of warfare was a strategic defense against bellicose nations. At age nineteen, Kossola was undergoing initiation for marriage. But these rites would never be realized. It was 1860, and the world Kossola knew was coming to an abrupt end.

TRANS-ATLANTIC TRAFFICKING

By the mid-nineteenth century, the Atlantic world had already penetrated the African hinterland. And although Britain had abolished the international trafficking of African peoples, or what is typically referred to as "the trans-Atlantic slave trade," in 1807, and although the United States had followed suit in 1808, European and

American ships were still finding their way to ports along the West African coast to conduct what was now deemed "illegitimate trade." Laws had been passed and treaties had been signed, but half a century later, the deportation of Africans out of Africa and into the Americas continued. France and the United States had joined forces with British efforts to suppress the traffic. However, it was a largely British-led effort, and the US patrols proved to be ambivalent and not infrequently at cross-purposes with the abolitionist agenda.[6]

Habituated to the lucrative enterprise of trafficking, and encouraged by the relative ease with which they could find buyers for their captives, Africans opposed to ending the traffic persisted in the enterprise. The Fon of Dahomey was foremost among those African peoples who resisted the suppression. Not only was the internal enslavement of their prisoners perceived as essential to their traditions and customs, the external sell of their prisoners afforded their kingdom wealth and political dominance. To maintain a sufficient "slave supply," the king of Dahomey instigated wars and led raids with the sole purpose of filling the royal stockade.

King Ghezo of Dahomey renounced his 1852 treaty to abolish the traffic and by 1857 had resumed his wars and raids. Reports of his activities had reached the newspapers of Mobile, Alabama. A November 9, 1858, article announced that "the King of Dahomey was driving a brisk trade at Ouidah."[7] This article caught the attention of Timothy Meaher, a "slaveholder" who, like many proslavery Americans, wanted to maintain the trans-Atlantic traffic. In defiance of constitutional law, Meaher decided to import Africans illegally into the country and enslave them. In conspiracy with Meaher, William Foster, who built the *Clotilda*, outfitted the ship for transport of the "contraband cargo." In July 1860, he navigated toward the Bight of Benin. After six weeks of surviving storms and avoiding being overtaken by ships patrolling the waters, Foster anchored the *Clotilda* at the port of Ouidah.

BARRACOON

From 1801 to 1866, an estimated 3,873,600 Africans were exchanged for gold, guns, and other European and American merchandise. Of that number, approximately 444,700 were deported from the Bight of Benin, which was controlled by Dahomey.[8] During the period from 1851 to 1860, approximately 22,500 Africans were exported. And of that number, 110 were taken aboard the *Clotilda* at Ouidah. Kossola was among them—a transaction between Foster and King Glèlè. In 1859, King Ghezo was mortally shot while returning from one of his campaigns. His son Badohun had ascended to the throne. He was called Glèlè, which means "the ferocious Lion of the forest" or "terror in the bush."[9] To avenge his father's death, as well as to amass sacrificial bodies for certain imminent traditional ceremonies, Glèlè intensified the raiding campaigns. Under the pretext of having been insulted when the king of Bantè refused to yield to Glèlè's demands for corn and cattle, Glèlè sacked the town.

Kossola described to Hurston the mayhem that ensued in the predawn raid when his townspeople awoke to Dahomey's female warriors, who slaughtered them in their daze. Those who tried to escape through the eight gates that surrounded the town were beheaded by the male warriors who were posted there. Kossola recalled the horror of seeing decapitated heads hanging about the belts of the warriors, and how on the second day, the warriors stopped the march in order to smoke the heads. Through the clouds of smoke, he missed seeing the heads of his family and townspeople. "It is easy to see how few would have looked on that sight too closely," wrote a sympathetic Hurston.[10]

Along with a host of others taken as captives by the Dahomian warriors, the survivors of the Bantè massacre were "yoked by forked sticks and tied in a chain," then marched to the stockades at Abomey.[11] After three days, they were incarcerated in the barracoons at Ouidah, near the Bight of Benin. During the weeks of his existence in the barracoons, Kossola was bewildered and anxious about his fate. Before him was a thunderous and crashing ocean that he had never seen before. Behind him was everything he called home. There in the barracoon, as there in his Alabama home, Kossola was transfixed between two worlds, fully belonging to neither.

KOSSOLA, HURSTON, CHARLOTTE MASON, AND "BARRACOON"

In September 1927, Hurston had met and come under contract with Charlotte Osgood Mason, a patron to several Harlem Renaissance luminaries. Mason funded Hurston's return to Alabama for the extended interviews with Kossola, and she supported Hurston's research efforts while preparing *Barracoon* for publication. In a March 25, 1931, letter to Mason, Hurston writes that the work "is coming along well." She reported that she had to revise some passages, but that she was "within a few paragraphs of the end of the whole thing. Then for the final typing." She described the revisions and related her new research findings: "I found at the library an actual account of the raid as

Kossula said that it happened. Also the tribe name. It was not on the maps because the entire tribe was wiped out by the Dahomey troops. The king who conquered them preserved carefully the skull of Kossula's king as a most worthy foe."[12]

Hurston and Mason conversed about the potential publication of *Barracoon* over a period of years. In her desire to see Hurston financially independent, Mason encouraged Hurston to prepare *Barracoon*, as well as the material that would become *Mules and Men*, for publication. Charlotte Mason considered herself not only a patron to black writers and artists, but also a guardian of black folklore. She believed it her duty to protect it from those whites who, having "no more interesting things to investigate among themselves," were grabbing "in every direction material that by right belongs entirely to another race." Following the suggestions of Mason and Alain Locke, Hurston advised Kossola and his family "to avoid talking with other folklore collectors—white ones, no doubt—who he and Godmother felt 'should be kept entirely away not only from the project in hand but from this entire movement for the rediscovery of our folk material.'"[13]

Mason's support of Hurston's efforts with *Barracoon* extended to monetary contributions to Kossola's welfare. Mason and Kossola would eventually communicate directly with each other, and Kossola would come to consider Mason a "dear friend." As one letter suggests, Kossola was struggling financially. It had come to Mason's attention that Kossola had used excerpts from his copy of Hurston's narrative to gain financial compensation from local newspapers. Kossola dictated a letter to Mason in response to her concern:

> Dear friend you may have seen in the papers about my History. But this has been over three years since I has let anyone take it off to copy from it. I only did that so they would help me. But there is no one did for me as you has. The lord will Bless you and will give you a long Life. Where there's no more parting, yours in Christ. Cudjo Lewis.[14]

As Mason was protective of Hurston's professional interests, both women remained concerned about Kossola's welfare. Having discovered that Kossola was not receiving money that Mason had mailed to him, Hurston looked into the matter. She updated Mason accordingly:

> I have written to Claudia Thornton to check up on Kossula and all about things. I have also asked the Post Office at Plateau to check any letters coming to Cudjoe Lewis from New York.[15]

As Hurston checked on Kossola, she continued revising the manuscript. "Second writing of Kossula all done and about typed," she wrote Mason on January 12, 1931. On April 18, she was enthusiastic: "At last 'Barracoon' is ready for your eyes."[16] Appreciative of Mason's support, Hurston dedicated the book to her and began submitting it to publishers. In September 1931, she contemplated Viking's proposal: "The Viking press again asks for the Life of Kossula, but in language rather than dialect. It lies here and I know your mind about that and so I do not answer them except with your tongue."[17] The dialect was a vital and authenticating feature of the narrative. Hurston would not submit to such revision. Perhaps, as Langston Hughes wrote in *The Big Sea*, the Negro was "no longer in vogue," and publishers like Boni and Viking were unwilling to take risks on "Negro material" during the Great Depression.[18]

THE GRIOT

There seems to be a note of disappointment in the historian Sylviane Diouf's revelation that Hurston submitted *Barracoon* to various publishers, "but it never found a taker, and has still not been published."[19] Hurston's manuscript is an invaluable historical document, as Diouf points out, and an extraordinary literary achievement as well, despite the fact that it found no takers during her lifetime. In it, Zora Neale Hurston found a way to produce a written text that maintains the orality of the spoken word. And she did so without imposing herself in the narrative, creating what some scholars classify as *orature*. Contrary to the literary biographer Robert Hemenway's dismissal of *Barracoon* as Hurston's re-creation of Kossola's experience, the scholar Lynda Hill writes that "through a deliberate act of suppression, she resists presenting her own point of view in a natural, or naturalistic, way and allows Kossula 'to tell his story in his own way.'"[20]

Zora Neale Hurston was not only committed to collecting artifacts of African American folk culture, she was also adamant about their authentic presentation. Even as she rejected the objective-observer stance of Western scientific inquiry for a participant-observer stance, Hurston still incorporated standard features of the ethnographic and folklore-collecting processes within her methodology. Adopting the participant-observer stance is what allowed her to collect folklore "like a new broom."[21] As Hill points out, Hurston was simultaneously working and learning, which meant, ultimately, that she was not just mirroring her mentors, but coming into her own.

Embedded in the narrative of *Barracoon* are those aspects of ethnography and folklore collecting that reveal Hurston's methodology and authenticate Kossola's story as his own, rather than as a fiction of Hurston's imagination. The story, in the main, is told from Kossola's first-person point of view. Hurston transcribes Kossola's story, using his vernacular diction, spelling his words as she hears them pronounced. Sentences follow his syntactical rhythms and maintain his idiomatic expressions and repetitive phrases. Hurston's methods respect Kossola's own storytelling sensibility; it is one that is "rooted 'in African soil.'" "It would be hard to make the case that she entirely invented Kossula's language and, consequently, his emerging persona," comments Hill.[22] And it would be an equally hard case to make that she created the life events chronicled in Kossola's story.

Even as Hurston has her own idea about how a story is to be told, Kossola has his. Hurston is initially impatient with Kossola's talk about his father and grandfather, for instance. But Kossola's proverbial wisdom adjusts her attitude: "Where is de house where de mouse is de leader?"[23]

Hurston complained in *Dust Tracks on a Road* of Kossola's reticence. Yet her patience in getting his story is quite apparent in the narrative. She is persistent in her returning to his home even when Kossola petulantly sends her away. He doesn't always talk when she comes, but rather chooses to tend his garden or repair his fence. And sometimes her time with him is spent driving Kossola into town. Sometimes he is lost in his memories.

Recording such moments within the body of the narrative not only structures the overall narrative flow of events but reveals the behavioral patterns of her informant. As Hurston is not just an observer, she fully participates in the process of "helping Kossula to tell his story." "In writing his story," says Hill, "Hurston does not romanticize or in any way imply that ideals such as self-fulfillment or fully realized self-expression could emerge from such suffering as Kossula has known. Hurston does not interpret his comments, except when she builds a transition from one interview to the next, in her footnotes, and at the end when she summarizes."[24] The story Hurston gathers is presented in such a way that she, the interlocutor, all but disappears. The narrative space she creates for Kossola's unburdening is sacred. Rather than insert herself into the narrative as the learned and probing cultural anthropologist, the investigating ethnographer, or the authorial writer, Zora Neale Hurston, in her still listening, assumes the office of a priest. In this space, Oluale Kossola passes his story of epic proportion on to her.

Deborah G. Plant

Editor's Note

Zora Neale Hurston's introduction to *Barracoon* has been edited to align with the conventions of spelling, punctuation, grammar, and usage. Contemporary spelling and usage have also been applied to names and places. In composing the introduction to her work, Hurston made a good-faith effort to document the source material she used to set the context for the *Barracoon* narrative. As she states in her preface, "For historical data, I am indebted to the *Journal of Negro History*, and to the records of the Mobile Historical Society." She reiterates this acknowledgment in her introduction and alludes to the use of other "records." Hurston drew from Emma Langdon Roche's *Historic Sketches*, but she references this work indirectly, and her citation from this book, as well as the other sources she utilized, was inconsistent. Wherever there is a question regarding her use of paraphrase and direct quotation, I have revised the passage as a direct quote and have documented it accordingly.

Regarding the actual narrative, I have read the original typescript in relation to earlier typed and handwritten drafts to produce a definitive text. Minor edits to the text were made in relation to the mechanics of typography, for purposes of clarity, or in the correction of apparent typos. Otherwise, the text remains as Hurston left it. I have made notations in the endnotes to present explanations or to provide full bibliographic data for sources Hurston used in her own notes. Such explanatory entries are labeled "Editor's note" and are bracketed. All other notes are original to the manuscript. Hurston's citations and footnotes have likewise been edited to align with conventional documentation style.

D.G.P.

The "Door of No Return" at La Maison des Esclaves (House of Slaves) at Gorée Island in Senegal, West Africa. Above the entryway: "Lord, give my people, who have suffered so much, the strength to be great" (Joseph Ndiaye).

Barracoon

To

Charlotte Mason

My Godmother, and the one Mother of all
the primitives, who with the Gods in Space is
concerned about the hearts of the untaught

Preface

This is the life story of Cudjo Lewis, as told by himself. It makes no attempt to be a scientific document, but on the whole he is rather accurate. If he is a little hazy as to detail after sixty-seven years, he is certainly to be pardoned. The quotations from the works of travelers in Dahomey are set down, not to make this appear a thoroughly documented biography, but to emphasize his remarkable memory.

Three spellings of his nation are found: *Attako, Taccou,* and *Taccow.* But Lewis's pronunciation is probably correct. Therefore, I have used *Takkoi* throughout the work.

I was sent by a woman of tremendous understanding of primitive peoples to get this story. The thought back of the act was to set down essential truth rather than fact of detail, which is so often misleading. Therefore, he has been permitted to tell his story in his own way without the intrusion of interpretation.

For historical data, I am indebted to the *Journal of Negro History*, and to the records of the Mobile Historical Society.

Zora Neale Hurston
April 17, 1931

Introduction

The African slave trade is the most dramatic chapter in the story of human existence. Therefore a great literature has grown up about it. Innumerable books and papers have been written. These are supplemented by the vast lore that has been blown by the breath of inarticulate ones across the seas and lands of the world.

Those who justified slaving on various grounds have had their say. Among these are several slave runners who have boasted of their exploits in the contraband flesh. Those who stood aloof in loathing have cried out against it in lengthy volumes.

All the talk, printed and spoken, has had to do with ships and rations; with sail and weather; with ruses and piracy and balls between wind and water; with native kings and bargains sharp and sinful on both sides; with tribal wars and slave factories and red massacres and all the machinations necessary to stock a barracoon with African youth on the first leg of their journey from humanity to cattle; with storing and feeding and starvation and suffocation and pestilence and death; with slave ship stenches and mutinies of crew and cargo; with the jettying of cargoes before the guns of British cruisers; with auction blocks and sales and profits and losses.

All these words from the seller, but not one word from the sold. The Kings and Captains whose words moved ships. But not one word from the cargo. The thoughts of the "black ivory," the "coin of Africa," had no market value. Africa's ambassadors to the New World have come and worked and died, and left their spoor, but no recorded thought.

Of all the millions transported from Africa to the Americas, only one man is left. He is called Cudjo Lewis and is living at present at Plateau, Alabama, a suburb of Mobile. This is the story of this Cudjo.

I had met Cudjo Lewis for the first time in July 1927. I was sent by Dr. Franz Boas to get a firsthand report of the raid that had brought him to America and bondage, for Dr. Carter G. Woodson of the *Journal of Negro History*. I had talked with him in December of that same year and again in 1928. Thus, from Cudjo and from the records of the Mobile Historical Society, I had the story of the last load of slaves brought into the United States.

The four men responsible for this last deal in human flesh, before the surrender of Lee at Appomattox should end the 364 years of Western slave trading, were the three Meaher brothers and one Captain [William "Bill"] Foster. Jim, Tim, and Burns Meaher were natives of Maine. They had a mill and shipyard on the Alabama River at the mouth of Chickasabogue Creek (now called Three-Mile Creek) where they built swift vessels for blockade running, filibustering expeditions, and river trade. Captain Foster was associated with the Meahers in business. He was "born in Nova Scotia of English parents."[1]

There are various reasons given for this trip to the African coast in 1859, with the muttering thunder of secession heard from one end of the United States to the other. Some say that it was done as a prank to win a bet. That is doubtful. Perhaps they believed with many others that the abolitionists would never achieve their ends. Perhaps they merely thought of the probable profits of the voyage and so undertook it.

The *Clotilda* was the fastest boat in their possession, and she was the one selected to make the trip. Captain Foster seems to have been the actual owner of the vessel.[2] Perhaps that is the reason he sailed in command. The clearance papers state that she was sailing for the west coast for a cargo of red palm oil. Foster had a crew of Yankee sailors and sailed directly for Whydah [Ouidah], the slave port of Dahomey.

The *Clotilda* slipped away from Mobile as secretly as possible so as not to arouse the curiosity of the Government. It had a good voyage to within a short distance of the Cape Verde Islands. Then a hurricane struck and Captain Foster had to put in there for repairs.

While he was on dry-dock, his crew mutinied. They demanded more pay under the threat of informing a British man-of-war that was at hand.

Foster hurriedly promised the increase the sailors demanded. But his wife often told how he laughingly broke this promise when it was safe to do so. After the repairs had been made, he made presents to the Portuguese officials of shawls and other trinkets and sailed away unmolested.[3]

Soon he was safely anchored in the Gulf of Guinea, before Whydah. There being no harbor, ships must stand in open roadstead and the communications with shore are carried on by Kroo men in their surf boats.

Soon Captain Foster and his kegs of specie and trading goods were landed. "Six stalwart blacks" were delegated to meet him and conduct him into "the presence of a Prince of Dahomey," but he did not meet the king.[4]

Foster was borne in a hammock to the Prince, who received him seated on his stool of rank. He was gracious and hospitable, and had Foster shown "the sights of Whydah."[5] He was surrounded by evidence of great wealth, and Foster was impressed. He was particularly struck by a large square enclosure filled with thousands of snakes, which he was told had been collected for ceremonial purposes.

The Prince expressed regret that Foster had arrived a little too late to witness the Dahomey "Custom" in honor of trade (foreign, i.e., mostly slave trade); nevertheless, he found Foster's company so pleasant that he wished to make him a present. He therefore desired Foster to look about him and select a person, "one that the 'superior wisdom and exalted taste' of Foster designated the finest specimen."[6] Foster looked about him and chose a young man named Gumpa; "Foster making this selection with the intention of flattering the Prince, to whom Gumpa was nearly related." This accounts for the one Dahoman in the cargo.[7]

The ceremonies over, Foster had "little trouble in procuring a cargo." The barracoons at Whydah were overflowing. "[I]t had long been a part of the traders' policy to instigate the tribes against each other," so that plenty of prisoners would be taken and "in this manner keep the markets stocked. News of the trade was often published in the papers." An excerpt from the *Mobile Register* of November 9, 1858, said: "'From the West coast of Africa we have advice dated September 21st. The quarreling of the tribes on Sierra Leone River rendered the aspect of things very unsatisfactory.'"[8]

Inciting was no longer necessary in Dahomey. The King of Dahomey had long ago concentrated all his resources on the providing of slaves for the foreign market. There was "a brisk trade in slaves at from fifty to sixty dollars apiece at Whydah. Immense numbers of Negroes were collected along the coast for export."[9]

King Ghezo maintained a standing army "of about 12,000, and of these 5000 are Amazons." The Dahoman year was divided into two parts—the wars and the festivals. "In the months of November or December the king commences his annual wars," and these wars were kept up until January or February.[10] These were never carried on for mere conquest. They were all forced upon the Dahomans from less powerful nations.

The King boasted that he never attacked a people unless they had not only insulted Dahomey, but his own people must ask him for a war against the aggressors for "three successive years." Then and then only would he let himself be persuaded to march forth and exterminate the insulting tribe. But there were so many insulting chiefs and kings that it kept the warriors of Dahomey, reluctant as they were, always upon the warpath. "[W]hole nation[s] are transported, exterminated, their name to be forgotten, except in the annual festival of their conquerors, when sycophants call the names of the vanquished countries to the remembrance of the victors."[11]

When the Dahoman king marched forth against a place, he concealed from his army "the name or the place against which he has brought them," "until within a day's march" of the goal. "Daylight is generally the time of onset, and every cunning, secrecy, and ingenuity is exercised to take the enemy by surprise." With or without resistance, "all the aged were decapitated on the spot" and the youth driven to the barracoons at Whydah.[12]

"On the return from war in January, the king resides at Cannah, and . . . 'makes a Fetish,'" that is, he "sacrifices largely and gives liberal presents" to the people and, "at the same time, purchases the prisoners and heads from his soldiers" of those slain in war. (The heads are always cut off and carried home. No warrior may boast of more enemies slain than he has heads to show for.) "[T]he slaves are then sold to the slave merchants, and their blood-money wasted in the ensuing Custom, Hwae-nooeewha, as the great annual feast is entitled in Dahoman parlance."[13]

The most important feast is "held in March, and called See-que-ah-hee," at which the king sacrifices many slaves and makes a great display of his wealth. There is a lesser festival in May or June "in honour of Trade" which is celebrated "with music, dancing, and singing." In July is celebrated the royal "salute to the Fetish of the Great Waters."[14]

Therefore, when Captain Foster arrived in May, the wars being just over for the year, he had a large collection to choose from. The people he chose had been in the stockade behind the great white house for less than a month. He selected 130, equal number of men and women, paid for them, got into his hammock and was conveyed across the shallow river to the beach, and was shot through the surf by the skillful Kroo boys and joined his ship. In other boats manipulated by the Kroo boys were his pieces of property.[15]

When 116 of the slaves had been brought aboard, Foster, up in the rigging, observing all the activities of the Port through his glasses, became alarmed. He saw all the Dahoman ships suddenly run up black flags.[16] "He hurried down and gave orders" to abandon the cargo not already on board, and to sail away with all speed. He says that the Dahomans were treacherously planning to recapture the cargo he had just bought and hold him for ransom. But the *Clotilda* was so expertly handled and her speed was so great that she sped away to safety with all ease.[17]

The next day he was chased by an English cruiser but escaped by pressing on sail. Nothing eventful happened until the 13th day when he ordered the cargo brought on deck so that they might regain the use of their limbs.

Though the space in the *Clotilda* greatly exceeded the usual space in most slavers, the blacks were cramped. "[T]he usual space in which the 'middle passage' was made was from two and a half to three feet in height."[18] It was about five feet in the *Clotilda*. However, the lack of action had numbed them.

"[O]n the twentieth day," Foster thought he saw a British cruiser on the horizon intercepting his course; he climbed to the mast with his glasses. Yes, there she was, sweeping on toward his course. He hurried down and gave orders for the slaves to be returned to the hold. Then he anchored and "lay until night," when he resumed his course.[19]

When Captain Foster reached American waters, the

slaves were put back in the hold. The ship lay hidden for three days "behind the islands in Mississippi Sound and near the lower end of Mobile Bay."

To make the hiding more secure, the *Clotilde* was dismasted. Then Foster got into a small boat, rowed by four sailors to go to the western shore of Mobile Bay, intending to send word to Meaher that the *Clotilde* had arrived. His approach was regarded with suspicion by some men ashore, and he was fired upon. Waving a white handkerchief their doubts were allayed and he offered fifty dollars for a conveyance which would take him to Mobile.[20]

"Captain Foster reached Mobile on a Sunday morning in August (1859)"; his return from the slave coast having been made in seventy days. "Arrangements had long been made that a tug should lie in readiness to go at a moment's notice down Mobile Bay to tow the *Clotilde* and her cargo to safety. When the news came, the tug's pilot was attending services at St. John's Church. Captain Jim Meaher and James Dennison—a Negro slave—hurried to the church" and called the pilot out. "The three hastened down to the wharf, and were soon aboard the tug." It proceeded down the bay, but waited till dark to approach the *Clotilda*.[21]

Finally, the tug was made fast to the *Clotilda* and "the trip up the bay was begun." *The last slave ship was at the end of its voyage*: "The tug avoided the Mobile River channel, slipped behind the light-house on Battery Gladden, into Spanish River. . . . As the *Clotilde* passed opposite Mobile the clock in the old Spanish tower struck eleven, and the watchman's voice floated over the city and across the marshes, 'Eleven o'clock and all's well.'"[22]

"The *Clotilde* was taken directly to Twelve-Mile Island—a lonely, weird place by night." There Captain Foster and the Meahers awaited the *R. B. Taney,* "named for Chief Justice Tainey" of the *Dred Scott* decision fame. Some say it was the *June* instead of the *Taney*.[23] "[L]ights were smothered, and in the darkness quickly and quietly" the captives were transferred from the *Clotilda* "to the steamboat [and] taken up the Alabama River to John Dabney's plantation below Mount Vernon." They were landed the next day, and left in charge of the slave, James Dennison.[24]

"At Twelve-Mile Island, the crew of Northern sailors again mutinied. Captain Foster, with a six shooter in each hand, went among them, discharged them, and ordered them to 'hit the grit and never be seen in Southern waters again.' They were placed aboard the tug" and carried to Mobile. One of the Meahers bought them tickets "and saw that they boarded a train for the North. The *Clotilde* was scuttled and fired, Captain Foster himself placed seven cords of light wood upon her. Her hull still lies in the marsh at the mouth of Bayou Corne and may be seen at low tide. Foster afterwards regretted her destruction as she was worth more than the ten Africans given him by the Meahers as his booty."[25]

The Africans were kept at Dabney's Place for eleven days: being only allowed to talk "in whispers" and being constantly moved from place to place.

At the end of the eleventh day clothes were brought to them and they were put aboard the steamer *Commodore* and carried to The Bend in Clark County, where the Alabama and the Tombigbee rivers meet and where Burns Meaher had a plantation.

There they were lodged each night under a wagon shed, and driven each morning before daybreak back into the swamp, where they remained until dark.[26]

"Meaher sent word secretly to those disposed to buy. They were piloted to the place of concealment by Jim Dennison. The Africans were placed in two long rows," men in one row, women in the other. Some couples were bought and taken to Selma.[27] The remainder were divided up among the Meahers and Foster, Captain Jim Meaher took thirty-two (sixteen couples); Captain Burns Meaher took ten Africans; Foster received ten; and Captain Tim Meaher took eight.[28] Finally, after a period of adjustment, the slaves were put to work. Before a year had passed, the war of Secession broke out. With the danger from interference from the Federal Government removed, all the Africans not sold to Selma were brought to the Meaher plantations at Magazine Point.

Nevertheless, the Meahers were tried in the federal courts 1860–61 and fined heavily for bringing in the Africans.[29]

The village that these Africans built after freedom came, they called "African Town." The town is now called Plateau, Alabama. The new name was bestowed upon it by the Mobile and Birmingham Railroad (now a part of the Southern Railroad System) built through [the town]. But still its dominant tone is African.

With these things already known to me, I once more sought the ancient house of the man called Cudjo. This singular man who says of himself, "*Edem etie ukum edem etie upar*": The tree of two woods, literally, two trees that have grown together. One part *ukum* (mahogany) and one part *upar* (ebony). He means to say,

"Partly a free man, partly free." The only man on earth who has in his heart the memory of his African home; the horrors of a slave raid; the barracoon; the Lenten tones of slavery; and who has sixty-seven years of freedom in a foreign land behind him.

How does one sleep with such memories beneath the pillow? How does a pagan live with a Christian God? How has the Nigerian "heathen" borne up under the process of civilization?

I was sent to ask.

I

It was summer when I went to talk with Cudjo so his door was standing wide open. But I knew he was somewhere about the house before I entered the yard, because I had found the gate unlocked. When Cudjo goes down into his back-field or away from home he locks his gate with an ingenious wooden peg of African invention.

I hailed him by his African name as I walked up the steps to his porch, and he looked up into my face as I stood in the door in surprise. He was eating his breakfast from a round enameled pan with his hands, in the fashion of his fatherland.

The surprise of seeing me halted his hand between pan and face. Then tears of joy welled up.

"Oh Lor', I know it *you* call my name. Nobody don't callee me my name from cross de water but you. You always callee me Kossula, jus' lak I in de Affica soil!"

I noted that another man sat eating with him and I wondered why. So I said, "I see you have company, Kossula."

"Yeah, I got to have somebody stay wid me. I been sick in de bed de five month. I needa somebody hand me some water. So I take dis man and he sleep here and take keer Cudjo. But I gittee well now."

In spite of the recent illness and the fact that his well had fallen in, I found Cudjo Lewis full of gleaming, good will. His garden was planted. There was deep shade under his China-berry tree and all was well.

He wanted to know a few things about New York and when I had answered him, he sat silently smoking. Finally, I told him I had come to talk with him. He removed his pipe from his mouth and smiled.

"I doan keer," he said, "I lakee have comp'ny come see me." Then the smile faded into a wretched weeping mask. "I so lonely. My wife she left me since de 1908. Cudjo all by hisself."

After a minute or two he remembered me and said contritely, "Excuse me. You didn't do nothin' to me. Cudjo feel so lonely, he can't help he cry sometime. Whut you want wid me?"

"First, I want to ask you how you feel today?"

Another muted silence. Then he said, "I thank God I on prayin' groun' and in a Bible country."

"But didn't you have a God back in Africa?" I asked him.

His head dropped between his hands and the tears sprung fresh. Seeing the anguish in his face, I regretted that I had come to worry this captive in a strange land. He read my face and said, "Excusee me I cry. I can't help it when I hear de name call. Oh, Lor'. I no see Afficky soil no mo'!"

Another long silence. Then, "How come you astee me ain' we had no God back dere in Afficky?"

"Because you said 'thank God you were on praying ground and in a Bible country.'"

"Yeah, in Afficky we always know dere was a God; he name Alahua, but po' Affickans we cain readee de Bible, so we doan know God got a Son. We ain' ignant—we jes doan know. Nobody doan tell us 'bout Adam eatee de apple, we didn't know de seven seals was sealee 'gainst us. Our parents doan tell us dat. Dey didn't tell us 'bout de first days. No, dass a right. We jes doan know. So dat whut you come astee me?"

I temporized. "Well, yes. I wanted to ask that, but I want to ask you many things. I want to know who you are and how you came to be a slave; and to what part of

Africa do you belong, and how you fared as a slave, and how you have managed as a free man?"

Again his head was bowed for a time. When he lifted his wet face again he murmured, "Thankee Jesus! Somebody come ast about Cudjo! I want tellee somebody who I is, so maybe dey go in de Afficky soil some day and callee my name and somebody dere say, 'Yeah, I know Kossula.' I want you everywhere you go to tell everybody whut Cudjo say, and how come I in Americky soil since de 1859 and never see my people no mo'. I can't talkee plain, you unnerstand me, but I calls it word by word for you so it won't be too crooked for you.

"My name, is not Cudjo Lewis. It Kossula. When I gittee in Americky soil, Mr. Jim Meaher he try callee my name, but it too long, you unnerstand me, so I say, 'Well, I yo' property?' He say, 'Yeah.' Den I say, 'You callee me Cudjo. Dat do.' But in Afficky soil my mama she name me Kossula.[1]

"My people, you unnerstand me, dey ain' got no ivory by de door. When it ivory from de elephant stand by de door, den dat a king, a ruler, you unnerstand me. My father neither his father don't rule nobody. De ole folks dat live two hund'ed year befo' I born don't tell me de father (remote ancestor) rule nobody.

"My people in Afficky, you unnerstand me, dey not

rich. Dass de truth, now. I not goin' tellee you my folks dey rich and come from high blood. Den when you go in de Afficky soil an' astee de people, dey say, 'Why Kossula over dere in Americky soil tellee de folks he rich?' I tellee you lak it tis. Now, dass right, ain' it?

"My father's father, you unnerstand me, he a officer of de king. He don't live in de compound wid us. Wherever de king go, he go, you unnerstand me. De king give him plenty land, and got plenty cows and goats and sheep. Now, dass right. Maybe after while he be a little chief, I doan know. But he die when I a lil boy. Whut he gointer be later on, dat doan reachee me.

"My grandpa, he a great man. I tellee you how he go."

I was afraid that Cudjo might go off on a tangent, so I cut in with, "But Kossula, I want to hear about *you* and how *you* lived in Africa."

He gave me a look full of scornful pity and asked, "Where is de house where de mouse is de leader? In de Affica soil I cain tellee you 'bout de son before I tellee you 'bout de father; and derefore, you unnerstand me, I cain talk about de man who is father (*et te*) till I tellee you bout de man who he father to him, (*et, te, te*, grandfather) now, dass right ain' it?

"My grandpa, you unnerstand me, he got de great big compound. He got plenty wives and chillun. His house, it is in de center de compound. In Affica soil de house of de husband it always in de center and de houses of de wives, dein in a circle round de house dey husband live in.

"He don't think hisself to marry wid so many women. No. In de Affica soil it de wife dat go findee him another wife.

"S'pose I in de Affica soil. Cudjo he been married for seven year for example. His wife say, 'Cudjo, I am growin' old. I tired. I will bring you another wife.'

"Before she speakee dat, she got de girl who he doan know in her mind. She a girl she think very nice. Maybe her husband never see her. Well, she go out in de market place, maybe in de public square. She see disa girl and astee de girl, 'You know Cudjo?' De girl tellee her, 'I have heard of him.' De wife say, 'Cudjo is good. He is kind. I like you to be his wife.' De girl say, 'Come with me to my papa and mama.'

"Dey go, you unnerstan me, to de girl's parents together. Dey astee her questions and she answeree for her husband. She astee dem questions too and if both sides satisfy wid one 'nother de girl's parents say, 'We give our daughter into yo' care. She ain' ours no mo'. You be good to her.'

"De wife she come back to Cudjo and makee de 'rangements. Cudjo got to pay de father for de girl. If she be a rich girl dat been in de fat-house long time, you unnerstand me, he got to pay two of everything for her. Two cow, two sheep, two goat, chickens, yam, maybe gold. De rich man, keepee his daughter in de fat-house long time. Sometime two year. She gittee de dinner in dere eight times a day and dey don't leavee her git in and out de bed by herself. De one whut keep de fat-house he lift dem in and out, so dey don't lose de fat.

"De man not so rich, he cain keep his girl dere long so she not so fat. So po' man don't send his daughter.

"Derefore, you unnerstand me, de man pay different price for different girl. If she de daughter of a po' family, or she been married before or somethin', he don't pay much for her.

"When de new wife come first to her husband compound she live in de house wid de old wife. She teach her what to do and how to take keer de husband. When she learn all dat, den she have a house by herself.

"When dey gittee ready to buildee de new house, de man takee de machete and chop de palm tree to mark de place where de house goin' be build. Den he throw down a cow and have plenty palm wine. Den all de people come and eatee de meat and drink de wine and stomp de place smooth and buildee de house.

"My grandpa, he buildee wife house many time.

"Some men in de Affica soil don't gittee no wife

'cause dey cain buy none. Dey ain' got nothing to give so a wife kin come to dem. Some got too many. When you hungry it is painful but when de belly too full it painful too.

"All de wives make food (*udia*) for de husband. All de men dey likee de fufu. He eatee de big calabash full to de top wid fufu, den my grandpa he lay down to sleep.

"De young wives (before they are old enough to take up the actual duties of wifehood) help put de husband to sleep. One makee wind for him wid de fan. Another one rub de head. Maybe one clean de hands and somebody look after de toe-nails. Den he sleepee and snore.

"Somebody stand guard before de door so nobody make noise and wakee him. Sometime de son of a slave in de compound makee too much noise. De man what stand guard ketchee him and takee him to my grandpa. He sit up and lookee at de boy so. Den he astee him, 'Whoever tellee you dat de mouse kin walk 'cross de roof of de mighty? Where is dat Portugee man? I swap you for tobacco! In de olden days, I walk on yo' skin!' (That is, I would kill you and make shoes from your hide.) 'I drink water from yo' skull.' (I would have killed you and used your head for a drinking cup.)

"My grandpa say dat, but he don't never astee de chief to sellee nobody to de Portugee. Some chief dey gittee mad when de slave talkee so sassy and don't do work lakee dey tell 'em. Den dey sell him to de Portugee. De chief throw orange under de table. Den he call the slave boy he goin' sell and say to de boy, 'Pick me de orange under de table.' De boy stoop under de table. De chief got a man standin dere. Maybe two. When de boy go under the table to gittee de orange, de chief say, 'Ketchee de bushman!' De men grabee de boy and sellee him.

"De chief he ain' always glad. One day de wife die. She still in de old wife's house and ain' never been no wife to the chief yet. She too young. Why she die, Cudjo doan know.

"When dey come to tell de chief his young wife dead, he go look. He slap his hand on his wrist. Den he scream in his fist and cry. He say, 'Yea! yea! yea! my wife dead. All my goods wasted. I pay big price for her. I fatten her and now she dead and I never sleep wid her once. Yea! yea! I lose so much! She dead and still a virgin! Yea! yea, tu yea! I have a great loss.'"

Cudjo looked out over his patch of pole-beans towards the house of his daughter-in-law. I waited for him to resume, but he just sat there not seeing me. I waited but not a sound. Presently he turned to the man sitting inside the house and said, "go fetchee me some cool water."

The man took the pail and went down the path between the rows of pole-beans to the well in the daughter-
in-law's yard. He returned and Kossula gulped down a
healthy cup-full from a home-made tin cup.

Then he sat and smoked his pipe in silence. Finally
he seemed to discover that I was still there. Then he
said brusquely, "Go leave me 'lone. Cudjo tired. Come
back tomorrow. Doan come in de mornin' 'cause den I
be in de garden. Come when it hot, den Cudjo sit in de
house."

So I left Cudjo sitting in his door with his bare feet
exposed to the cloud of mosquitoes that swarmed in
the shade of the inside of his house.

II

The King Arrives

The next day about noon I was again at Kossula's gate. I brought a gift this time. A basket of
Georgia peaches. He received me kindly and began to
eat the peaches at once. Mary and Martha, the twin
daughters of his granddaughter, wandered up to the
steps. The old man's love of these children was quite
evident. With glad eyes, he selected four of the finest
peaches and handed two to each little girl. He scolded
them on off to play with affectionate abuse. When they
were gone, he looked lovingly after them and pointed
to a little clump of sugarcane in the garden.

"See dat cane?" he asked.

I nodded that I did.

"Well, I plant dat cane. Tain much, but I grow dat so

when Martha and Mary come to me and say, 'Gran'pa I wantee some cane,' I go cut and give 'em.'"

There is a large peach tree in the yard that bears small but delicious clingstone peaches. They were beginning to ripen. The old man gave me one or two and put away one for each of the twins.

I was shown all over the gardens. Kossula was genial but not one word about himself fell from his lips.

So I went away and came again the following day. I brought another gift. A box of Bee Brand insect powder to burn in the house to drive out all the mosquitoes.

He was in a vocal mood and could scarcely wait until I set the powder burning to talk about his Affica.

So we settled on the porch and he talked. I reminded him that he had been telling me about the chief's losing a wife under such unfortunate circumstances and about his grandfather's compound.

"I doan fuhgittee nothin. I 'member everything since I de five year old.

"Yeah, my grandpa, he a officer of de king. He be wid de king everywhere he go, you unnerstan me.

"Derefore, you unnerstan me, one man he kill a leopard, well, de king doan keer 'bout he kill a leopard, but de law say dat when a man kill a leopard, he got to bring it to de king.

"De king doan want take de beast away from de man what kill it, you unnerstand me, but he got to take de big hairs (whiskers) dat grow round de mouth. Dey very poison, and de king doan want none de people to gittee kill. Some mens dey wicked, you unnerstand me, and dey take de hairs and make de poison. Derefore, you know, de king say when any man kill de leopard, he got to cover de head so no women kin see it and bring de leopard to de king.

"Den de drums go beat and callee all brave chiefs come discuss dis leopard dat been kill.

"De king he keep de head, de liver, de gall and de skin. Dat always belong to de king. It all makee different medicine. All de body, it he dried and makee more medicine too. But some tribe make fetish and eat de flesh, so dey eatee de medicine, you know.

"Derefore when a man kill de leopard and take de hairs before he let de king know he kill de leopard, dey kill that man. He a wicked man.

"One man you know, he kill a leopard. He cover de head and tie de body to a young tree. (Tied by the feet to a pole so that it could be carried.)

"Well de king call all de chiefs and they come lookee. Dey take off de cover from de head and de king look at de hairs. He see one hair it gone from de hole in

de face where it grow. All de chief dey lookee too. Dey
see de hair ain' dere. So dey call de man.

"De king say, 'Well, you killee dis beast?'

"De man say, 'Yeah, I kill him.'

"'How you kill dis leopard?'

"'Wid de spear, I kill him.'

"'Did you touch de head?'

"'No, I doan touchee de head at all. I only a common man and I know de head belong to de king. So I
doan touch it.'

"De king lookee at de head and lookee at de man.
He say, 'How is it dis beast got de hole for de hair but
one hair not dere. Tell me where de hair is. I see where
it pull out. Who is it dat you want to kill?'

"De man say, 'I doan want killee nobody. I ain'
touchee de hair. Dats de truth now. If I touchee de
hair, let *in-si-bi-di*' (that is, may I be turned over to
the executioner. *Insibidi* being the name for the executioner).

"Well dey search de man and find de hair. Den dey
try him. All day dey talk palaver. So nex' day dey find
him guilty. So dey say he got to die. He a wicked man
what speck to killee somebody wid de hair.

"Derefore, you unnerstand me, dey tie him by de
left foot and wait for *aku-ire-usen* (King's day, or great

day, all executions being saved for this day, though a few are executed on the Queen's day) den dey takee him to de place of sacrifice.

"De king come wid his seat and all de chiefs bring dey stool too. Dey seatee deyself and de drum beat. It speak wid de voice of de king. Den three *insibidi* come in de place and dance. One have a mouth-piece dat rattle. He shake de mouth-piece dat rattle. He shake de mouth-piece and sing.

"What he sing? Cudjo goin' tellee you:

"'On a great day like this, we kill de
One dat is evil
On a day like this we kill de bad one
Who would command the poison one
from the leopard to kill us.
On a great day like this we kill him
Who would kill the innocent?'

"He dance some more wid de drum and de other two dancee wid him. Den he sing some more:

"'A great knife dat eats no other blood but
    human blood.
Let it killee him.

It a great knife—it feed de earth
A great knife dat eats no other blood but
   human blood.'

"Dey dances some mo' when de king makee de sign, dey dance up to de man where he tied at and wid one lick, choppee de head off. De head fall to de ground and de mouth work so—it open and shut many time. But quick, they put a piece of de stick from de banana tree in de mouth. Den dey kin open de jaw when dey gittee ready. If dey don't do dat, de jaw close and dey cain git it open no mo'.

"De body of de man, dey bury it in de ground. De head, dey put it in de sacrifice place wid de other heads.

"De king go back to his village, but de chief have court every day. All day somebody say to him, 'Dis man, touch my wife! Disa man commit adultery!'

"Everything be done open dere. Not so many secrets. When a man kills somebody dere, he be tried open an' all de boys and men in de village hear de trial.

"I doan know how come he done it, but one man killee anudder one wid de spear. So dey 'rested dat man an' tie his hands wid palm cord. Den dey pick up de dead man an' carry him to de public square, de market place, you unnerstan. Den dey send message by de drum to de king in de village where he at to come

set on de trial an' 'cide de case. In Afficky, you unnerstand, if somebody steal or commit adultery, de chief of
de village, he try him. But if a man killee somebody,
den dey send for de king an' he come an' 'cide de case.
Therefore, when dis man spear de udder one through
de breast, dey send word for de king to come.

"De ole folks, you unnerstand me, de wise ones, dey
go out in de woods and gittee leaves—dey know which
ones—an' mashee de leaves wid water. Den dey paint
de dead man all over wid dis so he doan spoil till de
king come. Maybe de king doan git dere till de next
day. When de king come, my grandfather, he come
wid him.

"Befo' anybody see de king, we know he is almost
dere, because we hear de drum. When a little chief
travel, he go quiet, but when de king go any place, you
unnerstand me, de drum go befo' to let de people know
de king come.

"Dat night everybody sit up wid de dead man, all
night, an' eat meat and drink palm wine and banana
beer. Late de next day, you unnerstand me, de king
come, wid de chiefs of de udder villages, to help him
'cide de case. So de chief of our village, he went out a
short way to meet de king. Den he put down and killee
cows and goats. It too late dat day to hold a trial, you
unnerstand me. So dey 'cide to hold it next day. So dey

did. De king, he takes a special seat dey bring for him an' de chiefs from de udder towns, dey sit on dey stool of rank in different places aroun' de square.

"De dead man is laying on de ground in de center where everybody see him. De man dat kill him, he tied where folks kin see him too. Derefore, dey try de man.

"Dey askee de man why he killee dis udder one. He say de man work juju against him so his chile died, an' his cows dey stay sick all de time. De king say, 'If this man work juju against you, why doan you tell de chief an' de headman of de village? Why doan you tell de king? Doan you know we got law for people dat work juju? You ain' supposed to kill de man.'

"So dey talk an' all the chiefs settin' round, dey askee him questions, too.

"In Afficky de law is de law an' no man cain make out he crazy lak here, an' get excusee from de law. If you kill anybody, you goin die, too. Dey goin' killee you. So de king say, 'I hear de evidence, but this man got no cause to killee dat udder one. Derefore he must die.'

"De man stand dere. He doan cry. He doan talk. He jes' look straight at de king. Den all de chiefs dey gettee round de king and dey talkee while an' nobody know what dey say but dem. Den all de chiefs, dey go back an' takee dere seats again. Den de drums begin to play. De big drum, Kata kumba, de drum dat speaks lak a man, it begins to talk. An' de man what is *insibidi*, he begin to dance. Dey lead de murderer out into de center of de square. De *insibidi* he dance. (Gesture.) And as he dance, he watch de eye of de king, an' de eye of all de chiefs. One man will give him de sign. Nobody know which one will give de sign. Dey 'cide dat when dey was whispering together.

"Derefore de executioner dance until he get de sign of de hand. Den he dance up to de murderer and touch his breast with the point of de machete. He dance away again an' de next time he touch de man's neck wid his knife. The third time dat he touch de man, other men rush out and seize the murderer an' take-a de palm cord and stretch him face to face upon de dead man, an' tie him tight so he cain move hisself.

"When de executioner touch de murderer wid his knife, dat is a sign dat he is dead already. So dey wrap de cord around his neck and around de neck of de dead man. Dey wrap de cord around his body an' around de body of de dead man. Dey wrap his arm an' de dead man's arm wid de same cord. His leg is wrapped as one wid de leg of de man he done killed. So dey leave him dere. His nose is tied to de nose of de dead man. His lips touch the lips of de corpse. So dey leave him.

"De king an' de chief talk palaver 'bout other things while dey watchee de struggles of de murderer.

"Sometime if he be a strong man, an' de person he kill be little, he manage to get up and go a little away wid de body, but if de corpse be heavy, he lay right dere till he die.

"If he cry for water, nobody pay no attention to him because he is dead since the machete first touch him. So dey say, 'How can a dead man want water?' If he cry to be cut loose, nobody pay attention to him. Dey say, 'How can a dead man want to be loose? De udder dead man doan cry. How come this man cry?' So dey leave him dere.

"But people watch until he die too. How long it take? Sometime he die next day. Sometime two or three days. He doan live long. People kin stand de smell of de horse, de cow and udder beasts, but no man kin stand de smell in his nostrils of a rotten man."

III

When dey try de man dat steal de leopard hair, it de time to cut grass, so it don't choke de corn. Before de grass be dry 'nough to burn, my grandpa he take sick in his compound. How come he take sick, Cudjo doan know. I a li'l boy and I doan know why he die.

"But Cudjo know his father takee him to de compound of his father. I didn't see him after he died. Dey bury him right away so no enemy come look down in his face and do his spirit harm. Dey bury him in de house. Dey dig up de clay floor and bury him. We say in de Affica soil, 'We live wid you while you alive, how come we cain live wid you after you die?' So, you know dey bury a man in his house.

"De coffin settin dere just lak he in dere. De people come fetchu presents and place dem in de coffin. De first wife she set at de head of de coffin. When somebody came she cry. She cry with a song. De other wives dey join in and cry wid her.

"When we come in, de chief wife of my grandpa got up from de head of de coffin and throw de veil off her face. De udder wives throw off de veil too. De chief wife she weepee very loud and said, 'It is forty years since he married me, and now you find me a widow. Only yesterday he was worried about his wives and chillun and here he lies today in need of nothing!'

"My father say, 'Oh de ground eats de best of everything.' Den he weepee too. De chief wife she cry some more and de udder wives cry and shake de voice, 'Aiai, Aiai, Aiai!'

"De chief wife say, 'He was a wonderful man.' Den my father say, 'Dat is true, de ground kin prove it.'

"Den we set ourselves on de floor and de wives cover up dere faces and gittee quiet.

"De men sorry he dead too. Dey come bring presents and lookee at de coffin. Dey drink palm wine and sing sorrowful for him a song. '*O todo ah wah n-law yah-lee, owrran k-nee ra ra k-nee ro ro.*'

"Den somebody else come and de chief wife she rise

and start de weepin agin. It very sad. Dey see de head of all de wives is shaved. Dey see de cover over de face. Derefore, you unnerstan me, everybody feel sad.

"De first wife, she cry and say:

"'How long since we were married?
And now we are nothing but a widow

De husband what know how to keep women
De husband what know how to prepare a house
De husband what know every secret of women
De husband what knows what is needed
And gives it without asking—

How long since we were married?
And now we're nothing but a widow.'

"Dey call my grandpa brave and praise-giving names. Den dey cry with another song:

"'Whoever shake de leaf of dat tree
(a sweet shrub)
We are still smelling it.
Whoever kill our husband,
We shall never forget.'

"De wives cry lak dat every time somebody come in. When nobody came dey set quiet. Two years they must be widow. One year, dey don't touchee water to de face. Dey washee it always wid tears. In de Affica soil de women grieve for dey husband lak dat, you unnerstand me.

"All day, all night de people come, and every time somebody come, de women cry."

Kossula got that remote look in his eyes and I knew he had withdrawn within himself.

I arose to go. "You going very soon today," he commented.

"Yes," I said, "I don't want to wear out my welcome. I want you to let me come and talk with you again."

"Oh, I doan keer you come see me. Cudjo lak have comp'ny. Now I go water de tater vines. You see kin you find ripe peach on de tree and gittee some take home."

I put the ladder in the tree and climbed up in easy reach of a cluster of pink peaches. He saw me to the gate and graciously said goodbye.

"Doan come back till de nexy week, now I need choppee grass in de garden."

IV

In the six days between my visits to Kossula I worried a little lest he deny himself to me. I had secured two Virginia hams on my trip south and when I appeared before him the following Thursday, I brought him one. He was delighted beyond his vocabulary, but I read his face and it was more than enough. The ham was for *him*. For *us* I brought a huge watermelon, right off the ice, so we cut it in half and we just ate from heart to rind as far as we were able.

Then it was necessary to walk it down so he showed me over the Old Landmark Baptist Church, at his very gate, where he is the sexton.

Watermelon, like too many other gorgeous things in life, is much too fleeting. We lightened our ballast and returned to the porch.

"Now, you want me to tellee you some mo' about what we do in de Affica soil? Well, you good to me. I doan keer, I tellee you somethin'. It too hot I work anyhow.

"My father he name O-lo-loo-ay. He not a rich man. He have three wives.

"My mama she name Ny-fond-lo-loo. She de second wife. Now dat's right. I no tellee you I de son of de chief wife. Dat ain' right. I de son of de second wife.

"My mama have one son befo' me so I her second child. She have four mo' chillun after me, but dat ain' all de chillun my father got. He got nine by de first wife and three by de third wife. When de guls marry dey like see how many chillun dey kin have for dey husband."

"Aren't there some barren women?" I asked.

"No, dey all git chillun by dey husband. If dey doan gittee de babies, dey go talk to de ole folks. Den de old ones go in de bush and gittee de leaves and make a tea and give the girl some to drink. Den dey gittee babies for dey husband. Sometimes a woman doan never gittee no baby, though. Cudjo doan know (why).

"In de compound I play games wid all de chillun my father got. (See appendix.) We wrassle wid one 'nother. We see which one kin run de fastest. We clam de palm tree wid coconut on it and we eatee dat, we go in de woods and hunt de pineapple and banana and we eatee dat too. Know how we find de fruits? By de smell.

"Sometimes our mama say we run play 'nough. Dey tell us 'Dat, dat do? Come set down and I tellee you stories 'bout de animals, when they talk lak folks.' Cudjo doan know de time when de animals talk lak folks. De ole folks, dey tell me dat. Cudjo like very much to listen."

I said, "I like to hear stories too. Do you remember any of the stories your mama told you?"

"Well," said Kossula, "I tellee you de story nexy time you come set wid me. Now I tellee you 'bout Cudjo when he a boy back in de Affica. (See appendix for stories.)

"One day de chief send word to de compound. He want see all de boys dat done see fourteen rainy seasons. Dat makee me very happy because I think he goin' send me to de army. I then almost fifteen rainy seasons old.

"But in de Affica soil dey teachee de boys long time befo' dey go in de army. Derefore, you unnerstand me, when de boy 'bout fourteen dey start train him for de war.

"Dey don't go fight right away. No, first dey got to know how to walk in de bush and see and not show theyself. Derefore, first de fathers (elders) takee de boys on journey to hunt. Sometime it go and come back befo' night. Sometime it two, three sleeps (nights).

"Dey got to learn de step on de ground (tracks). Dey got to know whether whut dey hunt turned this way or that way. Dey learn to breakee de branch and turn it so dey kin find de way back home. Dey got to knot de long leaf so de folks behind kin know to follow.

"De fathers teachee us to know a place for de house (a camp site) and how we must choppee bark of de biggest tree so somebody else whut go running (traveling) kin know it a good place to sleepee.

"Me make de hunt many time. We shoot de arrows from de bow. We chunkee spear we kill de beastes and fetchee dem home wid us.

"I so glad I goin' be a man and fight in de army lak my big brothers. I likee beatee de drum too.

"Dey teachee us to sing de war song. We sing when we walk in de bush and make lak we goin' fight de enemy. De drum talkee wid us when we sing de song, '*Ofu, ofu, tiggy, tiggy, tiggy, tiggy batim, ofu ofu, tiggy tiggy, tiggy, tiggy batim! Ofu batim en ko esse!*'

("When the day breaks the cock shall crow
When the day breaks the cock shall crow
When the day breaks the cock shall crow

When someone crosses our roof we shall tear
A nation down."

The actual meaning is, "When we get there we shall
make our demands and if we are crossed we shall tear
down the nation who defies us.")

"Every year dey teachee us mo' war. But de king,
Akia'on, say he doan go make no war.[1] He make us
strong so nobody doan make war on us. We know de
secret of de gates so when de enemy come and we don't
know dey come, we kin run hidee ourself in de bush,
den dey don't see nobody dey go 'way. Den we come
behind dem and fight till dey all dead.

"Four, five rainy seasons it keep on lak dat, den I
grow tall and big. I kin run in de bush all day and not
be tired."

Kossula ceased speaking and looked pointedly at his
melon rind. There was still lots of good red meat and a
quart or two of juice. I looked at mine. I had more meat
left than Kossula had. Nothing was left of the first in-
stallment, but a pleasant memory. So we lifted the half-
rinds to our knees and started all over again. The sun
was still hot so we did the job leisurely.

Watermelon halves having ends like everything else,
and a thorough watermelon eating being what it is, a
long over-stuffed silence fell on us.

When I was able to speak, somehow the name juju came into my mind, so I asked Kossula what he knew about it. He seemed reluctant to answer my question, but finally he did so.

"I tellee you whut I know about de juju. Whut de ole folks do in de juju house, I doan know. I can't tellee you dat. I too young yet. Dat doan reachee me. I know dat all de grown men dey go to de mountain once a year. It have something to do wid makin' de weather, but whut dey do dere, Cudjo doan know. Now, dat's right. I doan make out I know whut go on wid de grown folks. When I come away from Afficky I only a boy 19 year old. I have one initiation. A boy must go through many initiations before he become a man. I jus' initiate one time.

"One day I was in de market place when I see a pretty girl walk past me. She so pretty I follow her a little way, but I doan speak. We doan do dat in Afficky. But I likee her. One ole man, he saw me watchee de girl. He doan say nothin' to me, but he went to my father an' say, 'Your boy is about breakin' de corn. He is getting to be a man an' knows de secret of man. So put goats down or a cow an' let us fix a banquet for him.' So my father say, 'All right.'

"But first dey doan fix de banquet for me. Dey have in Afficky a small stick on a string an' when dey make it go 'round fast, it roar like de lion or de bull. Dey have three kinds. One, dey call it de 'he' one de 'she' and one dey call it de dog 'cause dey make it bark data way. (The bull-roarer.)

"No woman mus' hear dis thing; if she do, she die. So dey stay inside and shuttee de door tight.

"Dey put me in de initiation house. After a while I hear a great roaring outside de door an' dey say to me, 'Go see where dat is.' Soon's I went outside I doan hear it at de door no more. It sound way off in de bush. They tell me to go in de bush to hunt it. As soon as I go to de bush to find out whut it is, I hear it behind me. I hear it behind me, in front of me, everywhere, but I never find it. De men are playing wid me. Way after while, dey take me into de banquet an' tell me de secret of de thing dat make de sound.

"At de banquet dey make me sit an' listen wid respect. Dey tell me, 'You are jus' below us. You are not yet a man. All men are still fathers to you.'

"There is plenty of roast meat and wine at de banquet an' all de men dey pinchee my ear tight to teach me to keep de secrets. Den I get a peacock feather to wear. In Americky soil I see plenty wimmins wear de peacock feather, but dey doan know what dey do. In Afficky soil a boy got to gittee plenty secrets inside dat he doan talk 'fo' he gittee de peacock feather."

V

When I gittee de peacock feather, I stand round de place where de chief talk wid de wise men. I hope dey see Cudjo and think he a grown man. Maybe dey call me to de council. De fathers doan never call me but I likee very much to be dere and lissen when dey talk.

"I likee go in de market place too and see de pretty gals wid de gold bracelets on de arm from de hand to de elbow. Oh, dey look very fine to Cudjo when dey walkee dey sling de arm so and de bracelet ring. I lak hear dat—it sound so pretty.

"One day I see one girl I lak very much to marry, but I too young to take a wife. But I lak her. I think 'bout her all de time. Derefore I go home and say to my folks, 'Be keerful how you treat such and such a girl.'

"Dey look at me den dey go ask for de girl to be my wife when I git li'l older.

"One day derefore I in de market, three men come whut strange to us. Dey say dey from Dahomey and dey wantee see our king. De king say, 'Alright, he talk wid dem.'

"Dey say, 'You know de king of Dahomey?'

"Akia'on say, 'I have heard of him.'

"De men from Dahomey say, 'Den you know all de strong names he got. You know he got one name, *Tenge Makanfenkpar*, a rock, the finger nail cannot scratch it, see! You know dey speak 'bout him and say, "*Kini, kini, kini,* Lion of Lions." Some say, "A animal done cut its teeth, evil done enter into de bush." (The "bush," meaning the surrounding tribes who feel the sharpness of Dahomey's tooth.) Dis king send to you and say he wish to be kind. Derefore you must sendee him de half yo' crops. If you doan send it, he make war.' (See note 1.)

"Our King Akia'on say, 'Astee you' king did he ever hear de strong name of Akia'on? Dey call me Mouth of de leopard? That he take hold on, he never let go. Tell him de crops ain' mine. Dey belong to de people. I cain send and take de people crops to send to de king of Dahomey. He got plenty land. Let him stop makin' slave hunt on udder people and make his own crops.'

"De king of Dahomey doan lak dat message, but Akia'on so strong, he 'fraid to come make war. So he wait. (See note 2.)

"De king of Dahomey, you know, he got very rich ketchin slaves. He keep his army all de time making raids to grabee people to sell so de people of Dahomey doan have no time to raise gardens an' make food for deyselves. (See note 3.)

"Maybe de king of Dahomey never come make raid in Takkoi, but one traitor from Takkoi go in de Dahomey. He a very bad man and de king (of Takkoi) say, 'Leave this country.' Dat man want big honors in de army so he go straight in de Dahomey and say to de king, 'I show you how to takee Takkoi.' He tellee dem de secret of de gates.

"Derefore, you unnerstand me, dey come make war, but we doan know dey come fight us. Dey march all night long and we in de bed sleep. We doan know nothin'.

"It bout daybreak when de folks dat sleep git wake wid de noise when de people of Dahomey breakee de Great Gate. I not woke yet. I still in bed. I hear de gate when dey break it. I hear de yell from de soldiers while dey choppee de gate. Derefore I jump out de bed and lookee. I see de great many soldiers wid French gun in de hand and de big knife. Dey got de women soldiers

too and dey run wid de big knife and make noise. Dey
ketch people and dey saw de neck lak dis wid de knife
den dey twist de head so and it come off de neck. Oh
Lor', Lor'!

"I see de people gittee kill so fast! De old ones dey
try run 'way from de house but dey dead by de door,
and de women soldiers got dey head. Oh, Lor'!"

Cudjo wept sorrowfully and crossed his arms on
his breast with the fingers touching his shoulders. His
mouth and eyes wide-open as if he still saw the grue-
some spectacle.

"Everybody dey run to de gates so dey kin hide
deyself in de bush, you unnerstand me. Some never
reachee de gate. De women soldier ketchee de young
ones and tie dem by de wrist. No man kin be so strong
lak de woman soldiers from de Dahomey. So dey cut
off de head. Some dey snatch de jaw-bone while de
people ain' dead. Oh Lor', Lor', Lor'! De poor folkses
wid dey bottom jaw tore off dey face! I runnee fast to
de gate but some de men from Dahomey dey dere too. I
runnee to de nexy gate but dey dere too. Dey surround
de whole town. Dey at all de eight gates.

"One gate lookee lak nobody dere so I make haste
and runnee towards de bush. But de man of Dahomey
dey dere too. Soon as I out de gate dey grabee me, and
tie de wrist. I beg dem, please lemme go back to my

mama, but dey don't pay whut I say no 'tenshun. Dey tie me wid de rest.

"While dey ketchin' me, de king of my country he come out de gate, and dey grabee him. They see he de king so dey very glad. Derefore, you unnerstand me, dey take him in de bush where de king of Dahomey wait wid some chiefs till Takkoi be destroy, when he see our king, he say to his soldiers, 'Bring me de word-changer' (public interpreter). When de word-changer came he say, 'Astee dis man why he put his weakness agin' de Lion of Dahomey?' De man changed de words for our king. Akia'on lissen. Den he say to de Dahomey king, 'Why don't you fight lak men? Why you doan come in de daytime so dat we could meet face to face?' De man changee de words so de king of Dahomey know what he say. Den de king of Dahomey say, 'Git in line to go to Dahomey so de nations kin see I conquer you and sell Akia'on in de barracoon.'

"Akia'on say, 'I ain' goin' to Dahomey. I born a king in Takkoi where my father and his fathers rule before I was born. Since I been a full man I rule. I die a king but I not be no slave.'

"De king of Dahomey askee him, 'You not goin' to Dahomey?'

"He tell him, 'No, he ain' goin' off de ground where he is de king.'

"De king of Dahomey doan say no mo'. He look at de soldier and point at de king. One woman soldier step up wid de machete and chop off de head of de king, and pick it off de ground and hand it to de king of Dahomey. (See note 4.)

"When I see de king dead, I try to 'scape from de soldiers. I try to make it to de bush, but all soldiers over-take me befo' I git dere. O Lor', Lor'! When I think 'bout dat time I try not to cry no mo'. My eyes dey stop cryin' but de tears runnee down inside me all de time. When de men pull me wid dem I call my mama name. I doan know where she is. I no see none my family. I doan know where dey is. I beg de men to let me go findee my folks. De soldiers say dey got no ears for cryin'. De king of Dahomey come to hunt slave to sell. So dey tie me in de line wid de rest.

"De sun it jus' rising.

"All day dey make us walk. De sun so hot!

"De king of Dahomey, he ride in de hammock and de chiefs wid him dey got hammock too. Po' me I walk. De men of Dahomey dey tie us in de line so nobody run off. In dey hand dey got de head of de people dey kill in Takkoi. Some got two, three head dey carry wid dem to Dahomey.

"I so sad for my home I ain' gittee hongry dat day, but I glad when we drink de water.

"Befo' de sun go down we come by a town. It got a red flag on de bush. De king of Dahomey send men wid de word-changer to de town and de chief come in de hammock and talk wid de king. Den he take down de red flag and hang a white flag. Whut dey say, Cudjo doan know. But he bring de king a present of yams and corn. De soldiers make fire and cook de grub and eatee. Den we march on. Every town de king send message.

"We sleepee on de ground dat night but de king and de chiefs hang dey hammock in de tree and sleepee in dem. Den nothin' doan harm dem on de ground. Po' me I sleepee on de ground and cry. I ain' used to no ground. I thinkee too 'bout my folks and I cry. All night I cry.

"When de sun rise we eat and march on to Dahomey. De king send word to every town we passee and de head-man come out. If dey got a red flag, dat mean dey 'gree dey ain' goin' pay no tax to de Dahomey. Dey say dey will fight. If it a white flag, dey pay to Dahomey whut dey astee dem. If it a black flag, dat mean dat de ruler is dead and de son not old 'nough to take de throne. In de Affica soil when dey see de black flag, dey doan bother. Dey know it be takin' advantage if dey make war when nobody in charge.

"De heads of de men of Dahomey got 'gin to smell very bad. Oh, Lor', I *wish* dey bury dem! I doan lak see my people head in de soldier hands; and de smell makee me so sick!

"De nexy day, dey make camp all day so dat de people kin smoke de heads so dey don't spoil no mo'. Oh Lor' Lor', Lor'! We got to set dere and see de heads of our people smokin' on de stick. We stay dere in dat place de nine days. Den we march on to de Dahomey soil."

Kossula was no longer on the porch with me. He was squatting about that fire in Dahomey. His face was twitching in abysmal pain. It was a horror mask. He had forgotten that I was there. He was thinking aloud and gazing into the dead faces in the smoke. His agony was so acute that he became inarticulate. He never noticed my preparation to leave him.

So I slipped away as quietly as possible and left him with his smoke pictures.

VI
Barracoon

It was Saturday when next I saw Cudjo. He was gracious but not too cordial. He picked me peaches and tried to get rid of me quickly, but I hung on. Finally, he said, "Didn't I tellee you not to come bother me on Sat'day? I got to clean de church. Tomorrow Sunday."

"But I came to help you, Kossula. You needn't talk if you don't want to."

"I thankee you come help me. I want you take me in de car in de Mobile. I gittee me some turnip seed to plant in de garden."

We hurriedly swept and dusted the church. Less than an hour later the Chevrolet had borne us to Mobile and back. I left him at his gate with a brief goodbye and tackled him again on Monday.

He was very warm this day. He glimmered and

glinted with light. I must first tell him about the nice white lady in New York who was interested in him.[1]

"I want you to write her a letter in de New York. Tell her Cudjo say a thousand time much oblige. I glad she send you astee me whut Cudjo do all de time."

I talked about the lady for a few minutes and my words evidently pleased him for he said, "I tellee you mo' 'bout Cudjo when he was in de Dahomey. I tellee you right. She good to me. You tell her Cudjo lak please her. She good to me and Cudjo lonely.

"Dey march us in de Dahomey and I see de house of de king. I cain tell all de towns we passee to git to de place where de king got his house, but I 'member we passee de place call Eko (Meko) and Ahjahshay. We go in de city where de king got his house and dey call it Lomey. (Either Abomey or Cannah.) De house de king live in hisself, you unnerstand me, it made out of skull bones. Maybe it not made out de skull, but it lookee dat way to Cudjo, oh Lor'. Dey got de white skull bone on de stick when dey come meet us, and de men whut march in front of us, dey got de fresh head high on de stick. De drum beat so much lookee lak de whole world is de drum dey beat on. Dat de way dey fetchee us into de place where de king got his house. (See note 5.)

"Dey placee us in de barracoon (stockade) and we

restee ourself. Dey give us something to eat, but not very much.

"We stay dere three days, den dey have a feast. Everybody sing and dance and beatee de drum. (1)

"We stay dere not many days, den dey march us to *esoku* (the sea). We passee a place call Budigree (Badigri) den we come in de place call Dwhydah. (It is called Whydah by the whites, but Dwhydah is the Nigerian pronunciation of the place.)[1]

"When we git in de place dey put us in a barracoon behind a big white house and dey feed us some rice.

"We stay dere in de barracoon three weeks. We see many ships in de sea, but we cain see so good 'cause de white house, it 'tween us and de sea.

"But Cudjo see de white men, and dass somethin' he ain' never seen befo'. In de Takkoi we hear de talk about de white man, but he doan come dere.

"De barracoon we in ain' de only slave pen at the place. Dey got plenty of dem but we doan know who de people in de other pens. Sometime we holler back and forth and find out where each other come from. But each nation in a barracoon by itself.

"We not so sad now, and we all young folks so we play game and clam up de side de barracoon so we see whut goin' on outside.

"When we dere three weeks a white man come in de barracoon wid two men of de Dahomey. One man, he a chief of Dahomey and de udder one his word-changer. Dey make everybody stand in a ring—'bout ten folkses in each ring. De men by dey self, de women by dey self. Den de white man lookee and lookee. He lookee hard at de skin and de feet and de legs and in de mouth. Den he choose. Every time he choose a man he choose a woman. Every time he take a woman he take a man, too. Derefore, you unnerstand me, he take one hunnard and thirty. Sixty-five men wid a woman for each man. Dass right.

"Den de white man go 'way. I think he go back in de white house. But de people of Dahomey come bring us lot of grub for us to eatee 'cause dey say we goin' leave dere. We eatee de big feast. Den we cry, we sad 'cause we doan want to leave the rest of our people in de barracoon. We all lonesome for our home. We doan know whut goin' become of us, we doan want to be put apart from one 'nother.

"But dey come and tie us in de line and lead us round de big white house. Den we see so many ships in de sea. Cudjo see many white men, too. Dey talking wid de officers of de Dahomey. We see de white man dat buy us. When he see us ready he say goodbye to de chief and gittee in his hammock and dey carry him cross de river. We walk behind and wade de water. It come up to de neck and Cudjo think once he goin' drown, but nobody drown and we come on de land by de sea. De shore it full of boats of de Many-costs. (See note 6.)

"De boats take something to de ships and fetch something way from de ships. Dey comin' and goin' all de time. Some boat got white man in it; some boat got po' Affican in it. De man dat buy us he git in a Kroo boat and go out to de ship.

"Dey takee de chain off us and placee us in de boats. Cudjo doan know how many boats take us out on de water to de ship. I in de last boat go out. Dey almost leavee me on de shore. But when I see my friend Keebie in de boat I want go wid him. So I holler and dey turn round and takee me.

"When we ready to leave de Kroo boat and go in de ship, de Many-costs snatch our country cloth off us. We try save our clothes, we ain' used to be without no clothes on. But dey snatch all off us. Dey say, 'You get plenty clothes where you goin'.' Oh Lor', I so shame! We come in de 'Merica soil naked and de people say we naked savage. Dey say we doan wear no clothes. Dey doan know de Many-costs snatch our clothes 'way from us. (See note 7.)

"Soon we git in de ship dey make us lay down in de dark. We stay dere thirteen days. Dey doan give us much to eat. Me so thirst! Dey give us a little bit of water twice a day. Oh Lor', Lor', we so thirst! De water taste sour. (Vinegar was usually added to the water to prevent scurvy—Canot.)[2]

"On de thirteenth day dey fetchee us on de deck. We so weak we ain' able to walk ourselves, so de crew take each one and walk 'round de deck till we git so we kin walk ourselves.

"We lookee and lookee and lookee and lookee and we doan see nothin' but water. Where we come from we doan know. Where we goin, we doan know.

"De boat we on called de *Clotilde*. Cudjo suffer so in dat ship. Oh Lor'! I so skeered on de sea! De water, you unnerstand me, it makee so much noise! It growl lak de thousand beastes in de bush. De wind got so much voice on de water. Oh Lor'! Sometime de ship way up in de sky. Sometimes it way down in de bottom of de sea. Dey say de sea was calm. Cudjo doan know, seem lak it move all de time. One day de color of de water change and we see some islands, but we doan come to de shore for seventy days.

"One day we see de color of de water change and dat night we stop by de land, but we don't git off de ship. Dey send us back down in de ship and de nexy mornin' dey bring us de green branch off de tree so we Afficans know we 'bout finish de journey.

"We been on de water seventy days and we spend some time layin' down in de ship till we tired, but many days we on de deck. Nobody ain' sick and nobody ain' dead.[3] Cap'n Bill Foster a good man. He don't 'buse us and treat us mean on de ship.

"Dey tell me it a Sunday us way down in de ship and tell us to keep quiet. Cap'n Bill Foster, you unnerstand me, he skeered de gov'ment folks in de Fort Monroe goin' ketchee de ship.

"When it night de ship move agin. Cudjo didn't know den whut dey do, but dey tell me dey towed de ship up de Spanish Creek to Twelve-Mile Island. Dey tookee us off de ship and we git on another ship. Den dey burn de Clotilde 'cause dey skeered de gov'ment goin' rest dem for fetchin' us 'way from Affica soil.

"First, dey 'vide us wid some clothes, den dey keer us up de Alabama River and hide us in de swamp. But de mosquitoes dey so bad dey 'bout to eat us up, so dey took us to Cap'n Burns Meaher's place and 'vide us up.

"Cap'n Tim Meaher, he tookee thirty-two of us. Cap'n Burns Meaher he tookee ten couples. Some dey sell up de river in de Bogue Chitto. Cap'n Bill Foster he tookee de eight couples and Cap'n Jim Meaher he gittee de rest.

"We very sorry to be parted from one 'nother. We cry for home. We took away from our people. We seventy days cross de water from de Affica soil, and now dey part us from one 'nother. Derefore we cry. We cain help but cry. So we sing:

"'Eh, yea ai yeah, La nah say wu
Ray ray ai yea, nah nah saho ru.'

"Our grief so heavy look lak we cain stand it. I think maybe I die in my sleep when I dream about my mama. Oh Lor'!"

Kossula sat silent for a moment. I saw the old sorrow seep away from his eyes and the present take its place. He looked about him for a moment and then said bluntly, "I tired talking now. You go home and come back. If I talkeed wid you all de time I cain makee no garden. You want know too much. You astee so many questions. Dat do, dat do (that will do, etc.), go on home."

I was far from being offended. I merely said, "Well when can I come again?"

"I send my grandson and letee you know, maybe tomorrow, maybe nexy week."

VII
Slavery

C ap'n Jim he tookee me. He make a place for us to sleepee underneath de house. Not on de ground, you unnerstand me. De house it high off de grounds and got de bricks underneath for de floor.

"Dey give us bed and bed cover, but tain 'nough to keepee us warm.

"Dey doan put us to work right away 'cause we doan unnerstand what dey say and how dey do. But de others show us how dey raisee de crop in de field. We astonish to see de mule behind de plow to pull.

"Cap'n Tim and Cap'n Burns Meaher workee dey folks hard. Dey got overseer wid de whip. One man try whippee one my country women and dey all jump on him and takee de whip 'way from him and lashee *him* wid it. He doan never try whip Affican women no mo'.

"De work very hard for us to do 'cause we ain' used to workee lak dat. But we doan grieve 'bout dat. We cry 'cause we slave. In night time we cry, we say we born and raised to be free people and now we slave. We doan know why we be bring 'way from our country to work lak dis. It strange to us. Everybody lookee at us strange. We want to talk wid de udder colored folkses but dey doan know whut we say. Some makee de fun at us.

"Cap'n Jim, he a good man. He not lak his brother, Cap'n Tim. He doan want his folks knock and beat all de time. He see my shoes gittee raggedy, you know, and he say, 'Cudjo, if dat de best shoes you got, I gittee you some mo'!' Now dass right. I no tellee lies. He work us hard, you unnerstand me, but he doan workee his folks lak his brother. Dey got de two plantation. One on de Tenesaw River and one on de Alabam River.

"Oh Lor'! I 'preciate dey free me! We doan have 'nough bed clothes. We workee so hard! De womens dey workee in de field too. We not in de field much. Cap'n Jim gottee five boats run from de Mobile to de Montgomery. Oh Lor'! I workee so hard! Every landing, you unnerstand me, I tote wood on de boat. Dey have de freight, you unnerstand me, and we have to tote dat, too. Oh Lor'! I so tired. No sleepee. De boat leak and we pumpee so hard! Dey ain' got no railing on de boat and in de night time if you doan watchee close you fall overboard and drown yo'self. Oh Lor'! I 'preciate dey free me.

"Every time de boat stopee at de landing, you unnerstand me, de overseer, de whippin' boss, he go down de gangplank and standee on de ground. De whip stickee in his belt. He holler, 'Hurry up, dere, you! Runnee fast! Can't you runnee no faster dan dat? You ain't got 'nough load! Hurry up!' He cutee you wid de whip if you ain' run fast 'nough to please him. If you doan git a big load, he hitee you too. Oh, Lor'! Oh, Lor'! Five year and de six months I slave. I workee so hard! Looky lak now I see all de landings. I callee all dem for you.

"De first landin' after de Mobile it de Twenty-One-Mile Bluff; de nexy it de Chestang; de nexy it de Mouth of de Tenesaw; den de Four Guns Shorter; den we pass Tombigbee; den de nexy it de Montgomery Hill; den de nexy it Choctaw Bluff; den de Gain Town; den Tay Creek; den Demopolis; den Clairborne; den Low Peachtree; den Upper Peachtree; den we come to de White Bluffs; den de Blue Bluffs; de nexy after dat it de Yellow Jacket. De river it is shallow dere sometime de boat hafter wait for de tide. De nexy after dat is Cahoba; den Selma; den Bear Landing; den Washington; den de last place it de Montgomery. I think I 'member dem, you unnerstand me, but I ain' been dere since 1865. Maybe I furgitee some. Doan lookee lak I never furgit. I work so hard and we ain' had nothin' to sleepee on but de floor. Sometime de bluff it so high we got to chunkee de wood down two three times fo' it git down where de river is. De steamboat didn't used to burnee de coal. It burnee de wood an' it usee so muchee wood!

"De war commences but we doan know 'bout it when it start: we see de white folks runnee up and down. Dey go in de Mobile. Dey come out on de plantation. Den somebody tell me de folkses way up in de North make de war so dey free us. I lak hear dat. Cudjo doan want to be no slave. But we wait and wait, we heard de guns shootee sometime but nobody don't come tell us we free. So we think maybe dey fight 'bout something else.

"De Yankees dey at Fort Morgan, you unnerstand me. Dey dere on account de war and dey doan let nothin' come passee dem. So po' folks, dey ain' gottee no coffee an' nothin'. We parchee de rice and makee de coffee. Den we ain' gottee no sugar, so we put de molassy in de coffee. Dat doan tastee so good, you unnerstand me, but nobody cain do nothin' 'bout it. Cap'n Jim Meaher send word he doan want us to starve, you unnerstand me, so he tell us to kill hogs. He say de hogs dey his and we his, and he doan wantee no dead folks. Derefo' you know we killee hogs when we cain gittee nothin'.

"When we at de plantation on Sunday we so glad we ain' gottee no work to do. So we dance lak in de Afficky soil. De American colored folks, you unnerstand me, dey say we savage and den dey laugh at us and doan come say nothin' to us. But Free George, you unnerstand me, he a colored man doan belong to nobody. His wife, you unnerstand me, she been free long time. So she cook for a Creole man and buy George 'cause she marry wid him. Free George, he come to us and tell us not to dance on Sunday. Den he tell us whut Sunday is. We doan know whut it is before. Nobody in Afficky soil doan tell us 'bout no Sunday. Den we doan dance no mo' on de Sunday.

"Know how we gittee free? Cudjo tellee you dat. De boat I on, it in de Mobile. We all on dere to go in de Montgomery, but Cap'n Jim Meaher, he not on de boat dat day. Cudjo doan know (why). I doan forgit. It April 12, 1865. De Yankee soldiers dey come down to de boat and eatee de mulberries off de trees close to de boat, you unnerstand me. Den dey see us on de boat and dey say 'Y'all can't stay dere no mo'. You free, you doan b'long to nobody no mo'.' Oh, Lor'! I so glad. We astee de soldiers where we goin'? Dey say dey doan know. Dey told us to go where we feel lak goin', we ain' no mo' slave.

"Thank de Lor'! I sho 'ppreciate dey free me. Some

de men dey on de steamboat in de Montgomery and dey got to come in de Mobile and unload de cargo. Den dey free too.

"We ain' got no trunk so we makee de bundles. We ain' got no house so somebody tellee us come sleepee in de section house. We done dat till we could gittee ourselves some place to go. Cudjo doan keer—he a free man den."

VIII
Freedom

After dey free us, you unnerstand me, we so glad, we makee de drum and beat it lak in de Affica soil. My countrymen come from Cap'n Burns Meaher Plantation where we is in de Magazine Point, so we be together.

"We glad we free, but den, you unnerstand me, we cain stay wid de folks what own us no mo'. Derefo' where we goin' live, we doan know. Some de folks from cross de water dey done marry and got de wife and chillun, you unnerstand me. Cudjo not marry yet. In de Affica soil when de man gottee de wife, he build de house so dey live together and derefo' de chillun come. So we want buildee de houses for ourselves, but we ain' got no lan'. Where we goin' buildee our houses?

"We meet together and we talk. We say we from

cross de water so we go back where we come from. So we say we work in slavery five year and de six months for nothin', now we work for money and gittee in de ship and go back to our country. We think Cap'n Meaher and Cap'n Foster dey ought take us back home. But we think we save money and buy de ticket ourselves. So we tell de women, 'Now we all want go back home. Somebody tell us it take lot of money to keer us back in de Affica soil. Derefo' we got to work hard and save de money. You must help too. You see fine clothes, you must not wish for dem.' De women tell us dey do all dey kin to get back in dey country, and dey tellee us, 'You see fine clothes, don't you wish for dem neither.'

"We work hard and try save our money. But it too much money we need. So we think we stay here.

"We see we ain' got no ruler. Nobody to be de father to de rest. We ain' got no king neither no chief lak in de Affica. We doan try get no king 'cause nobody among us ain' born no king. Dey tell us nobody doan have no king in 'Merica soil. Derefo' we make Gumpa de head. He a nobleman back in Dahomey. We ain' mad wid him 'cause de king of Dahomey 'stroy our king and sell us to de white man. He didn't do nothin' 'ginst us.

"Derefore we join ourselves together to live. But we say, 'We ain' in de Affica soil no mo' we ain' gottee no lan'.' Derefo' we talk together so we say, 'Dey bring us 'way from our soil and workee us hard de five year and six months. We go to Cap'n Tim and Cap'n Jim and dey give us de lan', so we makee houses for ourself.'

"Dey say, 'Cudjo, you always talkee good, so you go tell de white men and tellee dem whut de Affican say.'

"All de Afficans we workee hard, we gittee work in de saw mill and de powder mill. Some us work for de railroad. De women work too so dey kin help us. Dey doan work for de white folks. Dey raisee de garden and put de basket on de head and go in de Mobile and sell de vegetable, we makee de basket and de women sellee dem too.

"Derefo', you unnerstand me, it one day not long after dey tell me to speakee for lan' so we buildee our houses, Cudjo cuttin' timber for de mill. It a place where de school-house at now. Cap'n Tim Meaher come sit on de tree Cudjo just choppee down. I say, now is de time for Cudjo to speakee for his people. We want lan' so much I almost cry and derefo' I stoppee work and lookee and lookee at Cap'n Tim. He set on de tree choppin splinters wid his pocket knife. When he doan hear de axe on de tree no mo' he look up and see Cudjo standin' dere. Derefo' he astee me, 'Cudjo, what make you so sad?'

"I tell him, 'Cap'n Tim, I grieve for my home.'

"He say, 'But you got a good home, Cudjo.'

"Cudjo say, 'Cap'n Tim, how big is de Mobile?'

"'I doan know, Cudjo, I've never been to de four corners.'

"'Well, if you give Cudjo all de Mobile, dat railroad, and all de banks, Cudjo doan want it 'cause it ain' home. Cap'n Tim, you brought us from our country where we had lan'. You made us slave. Now dey make us free but we ain' got no country and we ain' got no lan'! Why doan you give us piece dis land so we kin buildee ourself a home?'

"Cap'n jump on his feet and say, 'Fool do you think I goin' give you property on top of property? I tookee good keer my slaves in slavery and derefo' I doan owe dem nothin? You doan belong to me now, why must I give you my lan'?'

"Cudjo tell Gumpa call de people together and he tell dem whut Cap'n Tim say. Dey say, 'Well we buy ourself a piece of lan'.'

"We workee hard and save, and eat molassee and bread and buy de land from de Meaher. Dey doan take off one five cent from de price for us. But we pay it all and take de lan'.

"We make Gumpa (African Peter) de head and Jaybee and Keebie de judges. Den we make laws how to behave ourselves. When anybody do wrong we make him 'pear befo' de judges and dey tellee him he got to stop doin' lak dat 'cause it doan look nice. We doan want nobody to steal, neither gittee drunk neither hurtee nobody. When we see a man drunk we say, 'Dere go de slave whut beat his master.' Dat mean he buy de whiskey. It belong to him and he oughter rule it, but it done got control of him. Now dass right, ain' it? When we speak to a man whut do wrong de nexy time he do dat, we whip him.

"Derefo' we buildee de houses on de lan' we buy after we 'vide it up. Cudjo take one acre and de half for his part. We doan pay nobody build our houses. We all go together and buildee de house for one 'nother. So den we gittee houses. Cudjo doan buildee no house at first 'cause he ain' got no wife.

"We call our village Affican Town. We say dat 'cause we want to go back in de Affica soil and we see we cain go. Derefo' we makee de Affica where dey fetch us. Gumpa say, 'My folks sell me and yo folks (Americans) buy me.' We here and we got to stay.

"Free George come help us all de time. De colored folks whut born here, dey pick at us all de time and call us ig'nant savage. But Free George de best friend de Afficans got. He tell us we ought gittee de religion and join de church. But we doan want be mixee wid de other folks what laught at us so we say we got plenty land and derefo' we kin build our own church. Derefo' we go together and buildee de Old Landmark Baptis' Church. It de first one round here."

Cudjo dismissed me by saying abruptly, "When you come tomorrow I like you take me down de bay so we gittee some crab."

IX
Marriage

He had on his battered hat when I drove up the next day. His rude walking stick was leaning against the door jamb. He picked it up and came on out to the car at once and we drove off. Without the least prompting he began to talk about his marriage.

"Abila, she a woman, you unnerstand me, from cross de water. Dey call her Seely in Americky soil. I want dis woman to be my wife. She ain' married, you unnerstand me, and I ain' gottee no wife yet. All de folks from my country dey got family.

"Whut did Cudjo say so dat dis woman know he want to marry her? I tellee you dat. I tellee you de truth how it was.

"One day Cudjo say to her, 'I likee you to be my wife. I ain' got nobody.'

"She say, 'Whut you want wid me?'

"'I wantee marry you.'

"'You think if I be yo' wife you kin take keer me?'

"'Yeah, I kin work for you. I ain' goin' to beat you.'

"I didn't say no more. We got married one month after we 'gree 'tween ourselves. We didn't had no wedding. Whether it was March or Christmas day, I doan remember now.

"Derefo', you know, we live together and we do all we kin to make happiness 'tween ourselves.

"Derefo', you unnerstand me, after me and my wife 'gree 'tween ourselves, we seekee religion and got converted. Den in de church dey tell us dat ain' right. We got to marry by license. In de Afficky soil, you unnerstand me, we ain' got no license. De man and de woman dey 'gree 'tween deyselves, den dey married and live together. We doan know nothin' 'bout dey have license over here in dis place. So den we gittee married by de license, but I doan love my wife no mo' wid de license than I love her befo' de license. She a good woman and I love her all de time.

"Me and my wife we have de six chillun together. Five boys and one girl. Oh, Lor'! Oh, Lor'! We so happy. Poor Cudjo! All de folks done left him now! I *so* lonely. We been married ten months when we have our first baby. We call him Yah-jimmy, just de same lak

we was in de Afficky soil. For Americky we call him
Aleck.

"In de Afficky we gottee one name, but in dis place
dey tell us we needee two names. One for de son, you
unnerstand me, and den one for de father. Derefo' I put
de name of my father O-lo-loo-ay to my name. But it
too long for de people to call it. It too crooked lak Kossula. So dey call me Cudjo Lewis.

"So you unnerstand me, we give our chillun two
names. One name because we not furgit our home; den
another name for de Americky soil so it won't be too
crooked to call.

"De nexy child we name him Ah-no-no-toe, den we
call him Jimmy. De nexy one name Poe-lee-Dah-oo.
He a boy, too. Den we have Ah-tenny-Ah and we call
him David. De las' boy we callee him my name, Cudjo,
but his Afficky name, it Fish-ee-ton. Den my wife have
one li'l girl and we call her Ee-bew-o-see, den we call
her Seely after her mama.

"All de time de chillun growin' de American folks
dey picks at dem and tell de Afficky people dey kill
folks and eatee de meat. Dey callee my chillun ig'nant
savage and make out dey kin to monkey.

"Derefo', you unnerstand me, my boys dey fight.
Dey got to fight all de time. Me and dey mama doan lak
to hear our chillun call savage. It hurtee dey feelings.

Derefo' dey fight. Dey fight hard. When dey whip de other boys, dey folks come to our house and tellee us, 'Yo' boys mighty bad, Cudjo. We 'fraid they goin' kill somebody.'

"Cudjo meetee de people at de gate and tellee dem, 'You see de rattlesnake in de woods?' Dey say, 'Yeah.' I say 'If you bother wid him, he bite you. If you know de snake killee you, why you bother wid him? Same way wid my boys, you unnerstan me. If you leavee my boys alone, dey not bother nobody!'

"But dey keep on. All de time, 'Aleck dis, Jimmy dat, Poe-lee dis an' t'other. David a bad boy. Cudjo fightee my son.' Nobody never say whut dey do to de Afficky savages. Dey say he ain' no Christian. Dey tell whut de savages do to dem, just lakee we ain' gottee no feelings to git hurtee.

"We Afficans try raise our chillun right. When dey say we ign'nant we go together and build de school house. Den de county send us a teacher. We Afficky men doan wait lak de other colored people till de white folks gittee ready to build us a school. We build one for ourself den astee de county to send us de teacher.

"Oh, Lor'! I love my chillun so much! I try so hard be good to our chillun. My baby, Seely, de only girl I got, she tookee sick in de bed. Oh, Lor'! I do anything to save her. We gittee de doctor. We gittee all de medi-

cine he tellee us tuh git. Oh, Lor'. I pray, I tell de Lor'
I do anything to save my baby life. She ain' but fifteen
year old. But she die. Oh, Lor'! Look on de gravestone
and see whut it say. August de 5th, 1893. She born 1878.
She doan have no time to live befo' she die. Her mama
take it so hard. I try tellee her not to cry, but I cry too.

"Dat de first time in de Americky soil dat death find
where my door is. But we from cross de water know
dat he come in de ship wid us. Derefo' when we buildee
our church, we buy de ground to bury ourselves. It on
de hill facin' de church door.

"We Christian people now, so we put our baby in
de coffin and dey take her in de church, and everybody come look down in her face. Dey sing, 'Shall We
Meet Beyond De River.' I been a member of de church
a long time now, and I know de words of de song wid
my mouth, but my heart it doan know dat. Derefo' I
sing inside me, '*O todo ah wah n-law yah-lee, owrran
k-nee ra ra k-nee ro ro.*'

"We bury her dere in de family lot. She lookee so
lonesome out dere by herself—she such a li'l girl, you
unnerstand me, dat I hurry and build de fence 'round
de grave so she have pertection.

"Nine year we hurtee inside 'bout our baby. Den we
git hurtee again. Somebody call hisself a deputy sheriff
kill de baby boy now. (Over)[1]

"He say he de law, but he doan come 'rest him. If my boy done something wrong, it his place come 'rest him lak a man. If he mad wid my Cudjo 'bout something den he oughter come fight him face to face lak a man. He doan come 'rest him lak no sheriff and he doan come fight him lak no man. He have words wid my boy, but he skeered face him. Derefo', you unnerstand me, he hidee hisself in de butcher wagon and when it gittee to my boy's store, Cudjo walk straight to talk business. Dis man, he hidin' hisself in de back of de wagon, an' shootee my boy. Oh, Lor'! He shootee my boy in de throat. He got no right shootee my boy. He make out he skeered my boy goin' shoot him and shootee my boy down in de store. Oh, Lor'! De people run come tellee me my boy hurtee. We tookee him home and lay him in de bed. De big hole in de neck. He try so hard to ketchee breath. Oh, Lor'! It hurtee me see my baby boy lak dat. It hurtee his mama so her breast swell up so. It make me cry 'cause it hurt Seely so much. She keep standin' at de foot of de bed, you unnerstand me, an' lookee all de time in his face. She keep telling him all de time, 'Cudjo, Cudjo, Cudjo, baby, put whip to yo' horse!'

"He hurtee so hard, but he answer her de best he kin, you unnerstand me. He tellee her, 'Mama, thass whut I been doin'!'

"Two days and two nights my boy lay in de bed wid de noise in de throat. His mama never leave him. She lookee at his face and tellee him, 'Put whip to yo' horse, baby.'

"He pray all he could. His mama pray. I pray so hard, but he die. I so sad I wish I could die in place of my Cudjo. Maybe, I doan pray right, you unnerstand me, 'cause he die while I was prayin' dat de Lor' spare my boy life.

"De man dat killee my boy, he de paster of Hay Chapel in Plateau today. I try forgive him. But Cudjo think that now he got religion, he ought to come and let me know his heart done change and beg Cudjo pardon for killin' my son.

"It only nine year since my girl die. Look lak I still hear de bell toll for her, when it toll again for my Fish-ee-ton. My po' Affican boy dat doan never see Afficky soil."

X

Kossula Learns About Law

Dey doan do nothin' to de man whut killee my son. He a deputy sheriff. I doan do nothin'. I a Christian man den. I a sick man, too. I done git hurtee by de train, you unnerstand me.

"Cudjo tell you how he git hurtee. I tellee you jes lak it were. Cudjo doan fuhgit it. It in March, you unnerstand me, and I makee de garden. It de 12th day of March, 1902.

"It a woman call me, you unnerstand me, to plow de field for her. She say, 'Cudjo, I like to git you plow de garden so I kin plant de sweet potatoes. I pay you.' I 'gree to dat.

"Derefo', you unnerstand me, I gittee up early de nexy mornin' and go plow de garden for her so den when I git through wid hers, I kin plant *my* garden.

I doan finish her garden 'cause my wife she callee me and scold me. She tell me, 'Cudjo, why you go workee hard lak dat befo' you eatee your breakfast? Dat ain' right. You goin' be sick. I gottee your breakfast ready long time. You come eat.'

"I go home wid Seely and eatee de breakfast. Den I think it goin' rain so I 'cide I plant my beans. Derefo', you unnerstand me, I tell my wife come to de field wid me and helpee me plant de beans.'

"She say, 'Cudjo, why you want me in de field? I cain plant no beans.'

"I tell her come on and drop de beans while I hill dem up. She come wid me and I show her how. After while she say, 'Cudjo, you doan need me drop no beans. You cain work 'thout no woman 'round you. You bringee me here for company.'

"I say, 'Thass right.'

"We ain' got 'nough beans. So I went to de market and astee de man for early beans, but he ain' got none. Derefo', you unnerstand me, I gittee my wife some meat and come home. Den I feedee my horse, an' my wife she cookin'. So den I curried de horse and it sprinkle rain. I stopee and study. I doan know if I go gittee mo' beans in de Mobile or if I wait. I 'cide to go fetch de beans. Derefo' I astee my wife to give me money.

"She put three dollars on de mantelpiece. I astee her,

'Seely, why you give me so muchee money? I doan need no three dollars.'

"She say, 'Spend whut you need and bringee de rest back. I know you ain' goin' wastee de money.'

"Den I hitch up de horse and go in de Mobile to git-tee de beans. Soon as I git de beans I turn back to go home.

"When I reachee de Government Street and Common, it de L and N railroad track dere, you unnerstand me. When I 'proach de track another rig goin' very slow in de middle of de road. So I make de passee de rig and jes' when I passee it an' git out on de track, de train rushee down on me. Oh, Lor'! I holler to dem to stop 'cause I dere on de track, but dey doan stop. It a switch engine, you unnerstand me. It rushee on and hittee de buggy an' knock me and hurtee my left side. Oh, Lor.' De horse gittee skeered and run away. My boy David find him nexy day and fetch him home.

"Somebody see de train hit me and hear me holler for dem to stop, dey come pickee me up and keer me to de doctor office. He givee me de morphine. A white lady on Government Street see me all hurtee and she see dat I took keer of. When I go home she send me a basket and visit me. She say de railroad ain' got no right smash up de buggy and hurtee me. I in de bed fourteen days. Dey broke three ribs. Dey ain' rung no bell. Dey ain' blow no whistle. She say she goin' see de company. Derefo' she go in de office of de L and N. De man in dere tellee her, 'We ain' goin' to do nothin'. It was daytime. Can't he see?'

"When I git able to git 'round de lady tellee me git me a lawyer and he makee de company pay me for hurtee me and 'stroy de buggy.

"Derefo', I go in de office of lawyer Clarke. He a big lawyer. Cudjo tell him, 'I ain' able to hire you. I want you to go to de company. I give you half.'

"De lawyer sue de company. De nexy year (1903) in January, dey send for me to 'pear in court. De judge say, 'De first case dis mornin' is Cudjo Lewis against de L an' N for $5,000.'

"I lookee hard. I say to myself, 'Who tell him dat? I didn't tell him I want $5,000.'

"De railroad lawyer say, 'We ain' goin' to give him nothin'.'

"Well, lawyer Clarke talk too. He say I all hurtee. I never be able to work no mo'. Dey pull off my short and lookee at de left side and de doctor say, 'No, Cudjo cain work no mo'.' Den lawyer Clarke say de railroad ought take keer me—done cripply me so bad.

"De railroad lawyer say dey not goin' give me nothin'. Dey say it broad daylight, ain' dis Cudjo got no eyes to see de great big train?

"Lawyer Clarke say, 'De train got a bell but dey din't ring it. Dey got a whistle, but dey din't blow it. De railroad track it layin' right cross de road. How kin de city of Mobile lettee de company makee de street dangerous and doan makee dem pay when people git hurtee on dey track? He talk a long time, den we all go out de courthouse to eatee de dinner.

"I tired, so I think I go home. I go gittee some flesh in de market to take home to Seely. David he stay in de court. He know de market I lak and derefo' he run ketchee me in de market 'fore I go home and tell me, 'Papa, de judge give you $650 from de company. De lawyer say you come tomorrow and gittee yo' money.'

"I doan go nexy day, but I send David. De lawyer say dat too soon. Come back nexy week. Well, I send and I send, but Cudjo doan gittee no money. In de 1904 de yellow fever come in de Mobile and lawyer Clarke take his wife and chillun and gittee on de train to run in de New York 'way from de fever, but he never gittee in de North. He die on de way. Cudjo never know whut come of de money. It always a hidden mystery how come I not killed when de train it standing over me. I thank God I alive today.

"De people see I ain' able to work no mo', so dey make me de sexton of de church."

XI

Cudjo's friends down the bay caught us a marvelous mess of blue crabs. We left these people late in the afternoon with many lingering exchanges of good wishes. On the way home we saw some excellent late melons in front of a store and bought two of them. I left one melon on his porch and took the other with me.

At the gate he called after me, "You come tomorrow and eatee de crab wid me. I lak you come keep me comp'ny!"

So the next day about noon, I was sitting on his steps, between the rain-barrels eating crabs. When the crabs were gone we talked.

"Let Cudjo tellee you 'bout our boy, David. He such a good boy. Cudjo doan fuhgit dat day. It Easter Saturday. He come home, you unnerstand me, and find me sweepin' de church. I been de sexton long time den. So he astee me, 'Papa, where mama?'

"I tell him, 'She in de house.'

"Derefo' he go in de house, you unnerstand me, and astee his mama what she goin' have for dinner. She tellee him she got de baked fish. He say, 'Oh I so glad we got baked fish. Gimme my dinner quick.' His mama astee him, 'When did you ever see me give you anything to eat befo' your Pa?' He say, 'Never.' She say, 'You takee yo' bath den maybe dat time yo' Pa here to eatee his dinner.'

De boy runnee back out to me and tell me make haste so he git something to eat. He hongry. I choppee de wood so he take de ax and choppee de wood hisself. I say, 'Go on, son, I ain' weak yet. I kin chop dis wood!' He say, 'No, I doan want you chop wood and I right here and strong.' Derefo' he choppee de wood and keer it in de house where his ma kin reachee it.

"Den we eat our dinner and David washee hisself and his mama put out de clean clothes for him to put on. He got on de unnershirt, but he ain' got on de top shirt. He ain' got no button on de unnershirt so me and his ma see de flesh. So I say, 'Son, fasten yo' clothes so yo' mama doan see de skin.' He lookee at hisself den he astee me, 'Who first saw me naked? My ma.' Den he laugh and put on de rest of de clothes. He say, 'Papa, mama, I go in de Mobile and gittee de laundry. Den I have clean shirts.'

"I astee him, 'How long befo' you come from town?' He say, 'Not long. Maybe I ketchee de same car back.'

"So he go leave de house.

"After while we hear somebody dey come laughing and talking. Seely say, 'David got a friend wid him.' I lookee to see who David got wid him, but it ain' David.

"Two men come tell me, 'Uncle Cudjo, yo' boy dead in Plateau.'

"I say, 'My boy not in Plateau. He in de Mobile.' Dey say, 'No, de train kill yo' boy in Plateau.'

"I tell dem, 'How kin de train kill my David in Plateau when he not dere? He gone in de Mobile to git his laundry. He be back after while.'

"Seely say, 'Go see, Cudjo. Maybe it not our boy. Go see who git killed.'

"Den I astee de men, 'Where dat man git killed you tellee me about?'

"Dey say, 'On de railroad track in Plateau.'

"Derefo', you unnerstan me, I go follow de people. Then I gittee to de place wid de big crowd stand 'round lookee.

"I go through de crowd and lookee. I see de body of a man by de telegraph pole. It ain' got no head. Somebody tell me, 'Thass yo' boy, Uncle Cudjo.' I say, 'No, it not my David.' He lay dere by de cross ties. One woman she face me and astee, 'Cudjo, which son of yours is dis?' and she pointee at de body. I tell her, 'Dis none of my son. My boy go in town and y'all tell me my boy dead.'

One Afficky man come and say, 'Cudjo, dass yo' boy.'

"I astee him, 'Is it? If dat my boy, where his head?' He show me de head. It on de other side de track. Den he lead me home.

"Somebody astee me, 'Cudjo, yo' boy dead. Must I toll de bell for you? You de sexton. You toll de bell for everybody else, you want me toll it for David?'

"I astee him, 'Why you want to toll de bell for David? He ain' dead.'

"De Afficky man told de people pick up de body and keer it home. So dey took de window shutter and lay de body on it and fetch it to Cudjo's gate. De gate, it too small, so dey lift it over de gate and place it on de porch. I so worried. I wishee so bad my David come back from town so de people stop sayin' dat my son on de shutter.

"When dey place de shutter on de porch, my wife she scream and fall out. De Afficky man say again, 'Cudjo, thass yo' boy.' I say, 'If thass my son, tell me where de head.' Dey brung it in a box and I lookee down in

David face. Den I say to de crowd, 'Git off my porch! Git out my yard!' Dey went. Den I fell down and open de shirt and pushee my hand in de bosom and feel de marks. And I know it my son. I tell dem toll de bell.

"My wife lookee at my face and she scream and scream and fell on de floor and cain raise herself up. I runnee out de place and fell on my face in de pine grove. Oh, Lor'! I stay dere. I hurtee so. It hurtee me so to hear Seely cry. Those who had come cross de water come to me. They say, 'Uncle Cudjo, come home. Yo' wife want you.' I say, 'Tell Seely doan holler no mo'. I cain stand it.'

"She promise me she won't holler if I come home. So I got back to de house. I astee de friend, 'Where de head.' He say, 'Dere yonder in de cracker box.' I tellee him, 'I want you to put it dere on de neck and fasten it so dat when people come in de mornin', dey won't know.'

"My friend he fasten de head so it doan look lak it cut off. Derefo' nexy day, when people come lookee in his face, he look jes lak he sleep.

"De bell toll again.

"Our house it very sad. Lookee lak all de family hurry to leave and go sleep on de hill.

"Poe-lee very mad 'cause de railroad kill his brother. He want me to sue de company. I astee him, 'Whut for?

We doan know de white folks law. Dey say dey doan pay you when dey hurtee you. De court say dey got to pay you de money. But dey ain' done it.' I very sad. Poe-lee very mad. He say de deputy kill his baby brother. Den de train kill David. He want to do something. But I ain' hold no malice. De Bible say not. Poe-lee say in Afficky soil it ain' lak in de Americky. He ain' been in de Afficky, you unnerstand me, but he hear what we tellee him and he think dat better dan where he at. Me and his mama try to talk to him and make him satisfy, but he doan want hear nothin. He say when he a boy, dey (the American Negro children) fight him and say he a savage. When he gittee a man dey cheat him. De train hurtee his papa and doan pay him. His brothers gittee kill. He doan laugh no mo'.

"Well, after while, you unnerstand me, one day he say he go ketchee some fish. Somebody see him go t'wards de Twelve Mile Creek. Lor', Lor'! He never come back."

There was a muted mournful pause, in which I could do nothing but wait with my eyes in the China-berry tree lest I appear indecently intrusive. Finally he came back to me.

"Excuse me I cain help it I cry. I lonesome for my boy. Cudjo know dey doan do in de Americky soil lak dey do cross de water, but I cain help dat. My boy gone. He ain' in de house and he ain' on de hill wid his mama. We both missee him. I doan know. Maybe dey kill my boy. It a hidden mystery. So many de folks dey hate my boy 'cause he lak his brothers. Dey doan let nobody 'buse dem lak dey dogs. Maybe he in de Afficky soil lak somebody say. Po' Cudjo lonesome for him, but Cudjo doan know.

"I try be very nice to Seely. She de mama, you unnerstand me, and derefo', you know she grieve so hard 'bout her chillun. I always try please her, you unnerstand me, but when we ain' got but two our chillun wid us, I cain stand see her look so lak she want cry all de time. We ain' got but one chile in de house wid us, 'cause Aleck, dat de oldest one, you unnerstand me, he married and live wid his wife. We buildee him a house right in de yard, jes lak in de Afficky soil.

"Look lak we ain' cry enough. We ain' through cryin. In de November our Jimmy come home and set round lak he doan feel good so I astee him, 'Son, you gittee sick? I doan want you runnin' to work when you doan feel good.' He say, 'Papa, tain nothin' wrong wid me. I doan feel so good.' But de nexy day, he come home sick and we putee him in de bed. I do all I kin and his mama stay up wid him all night long. We gittee de doctor and do whut he say, but our boy die. Oh Lor'! I good to my chillun! I want dey comp'ny, but looky lak dey lonesome for one 'nother. So dey hurry go sleep together in de graveyard. He die holdin' my hand.

"When we gittee back from de funeral, tain nobody in de house but me and Seely. De house was full, but now it empty. We old folks now and we know we ain' going have no mo' chillun. We so lonesome, but we know we cain gittee back de dead. When de spit goes from de mouf, it doan come back. When de earth eats, it doan give back. So we try to keep one 'nother comp'ny and be happy.

"I still sexton of de church. It growing to be a big church now. We call it de Old Landmark Baptis' Church, 'cause it de first one in Afficky Town. Dey done build mo' Baptis' churches now, but ours, it de first.

"My wife she help me all she kin. She doan lemme strain myself so I hurtee de side where de train hittee me.

"One day we plant, de nexy we reap so we go on."

Before I left I had Kossula's permission to photograph him.[1] But he forbade my coming back within three days. A cow had broken in his fence and was eating his potato-vines.

It was on a hot Saturday afternoon that I came to photograph Kossula.

"I'm glad you takee my picture. I want see how I look. Once long time ago somebody come take my picture but they never give me one. You give me one."

I agreed. He went inside to dress for the picture. When he came out I saw that he had put on his best suit but removed his shoes. "I want to look lak I in Affica, 'cause dat where I want to be," he explained.

He also asked to be photographed in the cemetery among the graves of his family.

XII
Alone

One night Seely wake up in de night and say, 'Cudjo wake up. I dream about our chillun. Look lak dey cold.' I tell her she think too much. Go back to sleep. It hurtee me, 'cause it a cold night in November in de 1908 and I 'member how Seely used to visit de chillun when dey was little to see dey got plenty quilts, so dey keep warm, you unnerstand me. De nexy day, she say 'Cudjo, come on less we go see our chilluns grave. So I say yes, but I try not take her 'cause I 'fraid she worry 'bout dem. So I go in de church and makee lak I busy so she furgitee de graveyard. When I come out de church, I don't see her nowhere, so I look cross de hill and I see her in de family lot. I see Seely goin' from one her chillun grave to de other, lak she cover dem up wid mo' quilts.

"De nexy week my wife lef' me. Cudjo doan know. She ain' been sick, but she die. She doan want to leave me. She cry 'cause she doan want me be lonesome. But she leave me and go where her chillun. Oh Lor'! Lor'! De wife she de eyes to de man's soul. How kin I see now, when I ain' gottee de eyes no mo'?

"De nexy month my Aleck he die. Den I jes lak I come from de Afficky soil. I got nobody but de daughter-in-law, Mary, and de grandchillun. I tellee her she my son's wife so she stay in de compound and she take de land when I go wid Seely and our chillun.

"Ole Charlie, he de oldest one come from de Afficky soil. One Sunday after my wife left me he come wid all de others dat come cross de water and say, 'Uncle Cudjo, make us a parable.'

"'Well den,' I say, 'You see Ole Charlie dere. S'pose he stop here on de way to church. He got de parasol 'cause he think it gwine rain when he leave de house. But he look at de sky and 'cide hit ain' gwine rain so he set it dere by de door an' go on to church. After de preachin' he go on home 'cause he think de parasol at Cudjo house. It safe. He say, "I git it nexy time I go dat way." When he come home he say to one de chillun, "Go to Cudjo house and tellee him I say sendee me my parasol."

"'De parasol it pretty. I likee keep dat one.' But I astee dem all, 'Is it right to keep de parasol?' Dey all say, 'No it belong to Charlie.'

"'Well,' I say, 'my wife, she b'long to God. He lef' her by my door.'

"I 'preciate my countrymen dey come see me when dey know I lonely. Another time dey come to me and say, 'Uncle Cudjo, make us another parable.'

"I bow my head in my hands, den I lift it up again. (Characteristic gesture when he begins a story.) Den I talk. 'I doan know—me and my wife, we been ridin. I think we go to Mt. Vernon. De conductor go to her and say, "Ole Lady, where you goin' get off?" She say, "Plateau."'

"'I look at her. I say, "How you say you goin' get off at Plateau? I thought you goin' to Mount Vernon wid me."'

"'She shake her head. She say, "I doan know. I jes know I git off at Plateau. I doan wanna leave you, but I got to git off at Plateau."'

"'De conductor blow once. He blow twice, and my wife she say, "Goodbye, Cudjo. I hate to leave you." But she git off at Plateau. De conductor come to me and astee, "Ole man, where you goin' git off?"'

"'I say, "Mount Vernon."'

"I travelling yet. When I git to Mount Vernon, I no talk to you no mo'."

I had spent two months with Kossula, who is called Cudjo, trying to find the answers to my questions. Some days we ate great quantities of clingstone peaches and talked. Sometimes we ate watermelon and talked. Once it was a huge mess of steamed crabs. Sometimes we just ate. Sometimes we just talked. At other times neither was possible, he just chased me away. He wanted to work in his garden or fix his fences. He couldn't be bothered. The present was too urgent to let the past intrude. But on the whole, he was glad to see me, and we became warm friends.

At the end the bond had become strong enough for him to wish to follow me to New York. It was a very sad morning in October when I said the final goodbye, and looked back the last time at the lonely figure that stood on the edge of the cliff that fronts the highway. He had come out to the front of his place that overhangs the Cochrane Highway that leads to the bridge of that name. He wanted to see the last of me. He had saved two peaches, the last he had found on his tree, for me.

When I crossed the bridge, I know he went back to his porch; to his house full of thoughts. To his memories of fat girls with ringing golden bracelets, his drums that speak the minds of men, to palm-nut cakes and bull-roarers, to his parables.

I am sure that he does not fear death. In spite of his long Christian fellowship, he is too deeply a pagan to fear death. But he is full of trembling awe before the altar of the past.

Appendix

TAKKOI OR ATTAKO—CHILDREN'S GAME

A memory test game played by two players. One player (A) the tester, squats facing the diagram which is drawn on the ground. The other player whose memory is to be tested squats with his back to the figure. A grain of corn is placed in each of the 3 circles between the lines. Each of the lines (1, 2, 3) has a name.

No 1 Ah Kinjaw Mah Kinney
No 2 Ah-bah jah le fon
No 3 Ah poon dacre ad meejie

A points at line 1 (at W) and B says, "Ah Kinjaw Mah Kinney." A points to line 2 and B says, "Ah-bah jah le fon." A goes on to line 3 and B says, "Ah poon dacre ad meejie." Then A points to circle No. 1 and B says, "Corn." A removes the grain of corn from the circle and goes back to line 1 at W. B recites the name again. A goes to line 2 and 3 as before then to circle 1. B says, "No corn." Then A points to circle 2 and B says, "Corn." A removes the corn from circle 2 and returns to line 1 (W), 2, and 3 and B gives the names as before. Then A goes to circle 1 and B says, "No corn." To circle 2 and B says, "No corn," to circle 3 and B says, "Corn." The corn is removed from circle 3 and A returns to line 1 at W and goes through the three lines and circles as before. Of course, if B remembers that there is no corn in any of the three circles, A then points to line 1 at X and B says, "Ah Kinjaw Mah Kinney" and A goes on to lines 2 and 3 and then on to circle 1 between X and Y and B says, "Corn." A removes the corn and returns to line at W and goes through the empty circles to lines at X and the empty circle. B says, "No corn" and A goes on to the next circle where B says, "Corn." The corn is removed then back to line 1 at W and the game keeps up until the twelve circles have been emptied of corn if B's memory is good enough.

Another game seems to be akin to both billiards and bowling. Three balls are racked up and the player stands off and knocks them down with seven balls in his hand. The top ball of the three must be hit last with the seventh thrown ball.

STORIES KOSSULA TOLD ME

There are no windows in Kossula's house. It was a cold day in December and the door was closed. The little light came from the pine knots in the fire place. It is crude, but suits his needs very well indeed. There are two pieces of iron slanting slightly upward in each inside wall of the fire place. It is an African idea transplanted to America. They are placed there to support the racks for drying fish. Kossula smokes a great deal and tamps his pipe quite often. All of his pipes have tops that he has made himself to keep the fire from falling out as he works. The pipe lids are just another of the evidences of the primitive, the self-reliance of the people who live outside the influence of machinery.

There is something in the iron pot bubbling away among the coals. We eat some of the stew and find it delicious. It is a sort of stew of all flesh shredded in some way.

Kossula lights his pipe again. "You want me to tell you story 'bout Afficky? I done fuhgit all dat. I been in Americky soil de sixty-nine year last Augus'. It been so long I have anybody talk wid, I fuhgit. You don't be mad wid Uncle Cudjo if he fuhgit, Baby? I wouldn't hurty yo' feelin' fuh nothin' in dis world."

I assure him that I can never be angry with him, no matter if he never remembers a word, but praying strongly within that he remembers. We sit for a long time in silence. I tell him a few stories, after giving him a chance to think, and he is delighted. Finally he turns eagerly towards me, his face alight.

"I gwine tell you disa story:—

"Tree men, you unnerstand me, dey agree dey ain' goin' tell one on de udder.

"One day dese tree men dey say, 'We ain' got no meat—less we go in de woods and fin' a cow and 'vide it up.'

"Dey hunt till dey fin' a fat one and dey kill hit. Dey all git roun' it. One say, 'I want a hind leg.' Other say, 'I want a hind leg.' Third one say, 'I want a hind leg.' (A beaming face is turned to me to see if I get the point that three men can't get a hind leg off of one cow. He is very happy that I appreciate the dilemma in the tale.) Dey 'gin fight and fight. One say, 'I killee you.' (Very expressive gesture of conflict.) Other say, 'I killee *you.*' (Very hearty laughter, the struggling gestures continue.) Dey fight till dey come to de highway and de officer see dem fightin', you unnerstand me, and he say, 'Lookee heah, whut y'all fightin' 'bout?'

"One de men he say, "'If you don't foolee me, I won't foolee you.'"

"He axed de other. He say, 'If you don't foolee me, I won't foolee you.' De third man he say de same thing, so de officer he go to de king an' say, 'I found tree men dey fight, but when I axee dem whut for dey fight, dey all say, "If you don't foolee me, I won't foolee you."'

"De king summons dem tuh 'pear befo' him and he say. He say, 'Whusa matta you tree men?' Dey all say same thing agin. (Hearty chuckling.) Den de king he say, 'Something dey do, dey doan wanna tell. Dey is men of strong friendship.' Den he give dem ten coats, ten shoes, ten of everything and sent dem off. Dey went back and 'vide de cow ekal."

Mirthy tears ran down the cheeks of Kossula and he shook with chuckles long after the tale was finished. But he could not be persuaded to tell another that day. "You come agin Tuesday, nexy week an' I tellee you somethin' if I think. But Uncle Cudjo gittin' ole. I been in de Mericky soil since 1859. I fuhgits."

On the Tuesday after the New Year, I found Cudjo in a backward-looking mood. He was with his departed family in the land to the west. He talked about his boys, he grew tearful over his wife.

"I so lonely. I los' my wife de 15 November 1908. We been together long time. I marry her Chris'mas day, 1865. She a good wife to me."

There was a long, feeling silence, then he turned to me and spoke, "Ole Charlie, he de oldest one come from Afficky, came one Sunday after my wife lef' me and say, 'Uncle Cudjo, make us a parable.'

"Den I axed dem, 'How many limbs God give de body so it kin be active?'

"Dey say six; two arms two feet two eyes.

"I say dey cut off de feet, he got hands to 'fend hisself. Dey cut off de hands he wiggle out de way when he see danger come. But when he lose de eye, den he can't see nothin' come upon him. He finish. My boys is my feet. My daughter is my hands. My wife she my eye. She left, Cudjo finish."

It was two o'clock, and Kossula excused himself that he might work on his fence before dark. "Come see me when tain cold."

Two days later I sat beside his fire in the windowless house, and watched him smoke until he was ready to speak. I told him a story or two and finally he glowed and stirred.

"It a man, you know, he got a son. Six men, you un-nerstand, dey follow him all de time. De long runnee, de ole man say, 'Son, dese men always in yo' house. You know whut six men do to you?'

"'Dey don't do nothin' to me,' dat whut de son say, an' always de seven men be together till he git grown, and de time come for him to marry.

"De ole man, he want to try dese six men. So when de son marry, he hide de girl an' den he take a ramma (ram) and he kill hit an' cut off de horns. He fix it an' make it look like de girl.

"Den he say to de boy, 'Go tell your friends dat you marry de girl las' night and she fell dead an' I don't want de king to know; an' dig a grave (he wants the friends to dig the grave) an' bury her. Perhaps she was too young an' never had know no man.'

"Well, de six men come to dig de grave, but only two stay to finish dig, an' four went spread de news, clean till it reached to de king.

"De king den sent for de ole man an' say to him, 'Yo son jus' married a girl. Where she?'

"'She at home,' de ole man say to de king, an' he say, 'Where yo' house? I wants to see.'

"De king goes wid him to de house an' he show him de girl. Den he say, 'Well, whut you bury in de hole?'

He say 'De ramma.' But de king want satisfied and he hafta dig up de grave and let de king see de ramma hisself. Den he tell de king how tis.

"'I aska my boy 'bout these six men and he say dey all right. All de time dey sleep an' eat an' go wid him. I want know dey friendship so I killa de ramma.'

"De king say, 'You have knowledge,' an' so he paid the two whut stay dig de grave an' don't say nothin' an' killed de four men whut talk an' betray dey friend."

The Monkey and the Camel

One day—I tellee disa one—de *uthucudum* (weasel) went up de melon tree to eatee hisself some fruit. De camel, he lak melon all de time; so when he see de weasel in de tree, he go aska him throw him some. De weasel throw *him* some, den he come down and go in his house.

De camel, he still wantee some more melon, so he wait. After while, de monkey he go to de melon tree to gittee him some too. De camel, he hurry up under de tree and say to de monkey, "Gimmee some de melons too," and de monkey throw him some.

Den he aska de monkey to throw him more and he eata dat, den he aska for more and more till de monkey he git tired. He want to come down from de tree an' go home to dinner wid his fruit, so he tell de camel he too greedy and if he want more melon, let him clam' de tree hisself and gittee some.

Dat make de camel mad so he say dat de monkey is a very moufy animal wid an ugly red behind, and ver' ugly nose.

Now de monkey he know dat his nose is ugly and he is very shamed for the camel to speaky 'bout it, so he say dat de camel is a creature widee no hindquarters.

De camel gittee so mad at dat, till he reach up de tree and grab de monkey and carry him off.

Well, after he walk a while, he meeta de rhinoceros and he aska, "Camel, how come you seize de monkey?"

De camel say, "Let him tell himself."

De monkey, "Well I was up de melon tree eatin' some fruit and de camel come 'long and aska me throw him some. I *did* throw him some, and more and more, but when I was tired and want to go home he says dat I am a beast wid ugly nostrils and sunken eyes, and I got very mad say dat de camel is a beast widout a rump and he done seized me and tote me off."

De rhinoceros said dat de monkey was wrong to speak of de camel so and told him not to let him go, so de camel carried him on.

After a while, dey met de leopard and said, "O camel, what makee you seize de monkey? Is he done you wrong?"

"Let him tellee you hisself what he done."

De monkey, "Well I was up de melon tree eatin' some fruit an' de camel he come up under de tree and aska me to throw him down some fruit. Well, I throw him some, den some more, den some more till I gittee very tired den I say he is a lazy animal dat worries other animals when dey go to git fruit, let him clam de melon tree hisself. Den he say I am a creature wid no manners and a red behind, and I say dat he is a beast wid no behind at all, and not enough tail to hide de place where his behind ought to be; den he grabee me and bring me here."

De leopard say dat de monkey was wrong to speak thus of the camel and that the camel must not let him go; so de camel carried him on.

After a while dey come to de house of de weasel, and he was sittin' outside de do'way. He seen de camel wid de monkey and he aska de camel, "O camel, how come you seize de monkey? Whut he done wrong?"

De camel say, "Let him tell it hisself."

De monkey say, "Well, I was up de melon tree gittin' fruit for my wife, and de camel come under de tree and aska me to throw him down some fruit, I done throw him some, den more, den more, till I was tired, and I said he was a greedy beast whose rump looked lak he been drinking kainya (a powerful laxative) and he grabee me and bring me here."

Now de weasel he feel sorry for de monkey and he know hisself dat de camel is worrysome under de fruit tree, so he set a while den he say, "I will be de judge 'twixt you two," and dey both say, "All right, you be de judge for us."

De first thing, he say, "You monkey, come set here on my right side, and you camel set here on my left whilst I decide de question."

Dey both done whut he say, and he set dere quiet for a while. Den he open his mouf, "O monkey, I sentence you, for speaking so to de camel, to leap up dat tree, whilst I run into my hole," and he done dat and de camel was lef' settin' where he was. After a while he went away.

Story of de Jonah

Whut you want me to talk, Jonah?

Who and whut kinda prophet is Jonah, I doan know. I couldn't tellee you dat.

God speakee unto Jonah, go tell Ninevah to turn to me 'cause they sins it come befo' me. Jonah say no, I ain' gwine. Jonah say well, being I here, he gointer torment me, I goin' git away from here.

So he went dere, you know, in de vessel ship to go to Joppy—dat a country, you know, where God ain' gwine bother him. Listen, Cudjo say so, he didn't know it, God is everywhere. And so he went onto de ship to go to Joppy and God lookee at him. God see Jonah in de vessel and so when he went to de vessel God *lookee* (gesture of a penetrating look) at him. He see de Jonah dere.

He *see* de Jonah dere, so God went to de east and (gesturing of unlocking and flinging wide a door) un-lockee de storm room, say to de storm, "Come out" (hand uplifted in a kingly commanding gesture) and de storm started. Den God went to de west, unlockee an-udder storm room. (Gesture.) "Come out! Come outa dere!" Den God went to de north, unlockee dat storm room, tell it to come out! Den he went to de south, un-lockee anudder storm room, and anudder storm in de south. All storms come meetee together! All storms comee meetee together, and de vessel can't go no where.

Now! Whut did de captain say? Dat whut I go tellee you now. De captain say, "Dats not de first time I go travel in de sea. Something wrong!" And de man say, "Captain, dere's a man in de boat and den he pay his fare." De captain say, "Where 'bouts is he?" Dey say, "He way down in de bottom of de boat." He say, "Go tell him to come here."

I goin' tell you whut de sailors say when dey went down in de bottom of de boat. I goin' tellee you whut dey say to Jonah. Dey say, "O sleeper, wake up from your sleep and call on your God, else we go perish in de sea!"

When he come to de captain on de deck he say, "Who you?" He (Jonah) say, "I'm a Hebrew, done run away from God." Captain say, "Whut must we do now so de sea kin become calm?" He say, "Heave me overboard." De captain say, "I ain' gwine do it till we draw de lot. We don't want be guilty of your blood."

Dey draw de lot and de lot fell on Jonah. Lookee here God prepare de whale right long sidee de ship wi' his mouf wide open (gesture). When dey throw him in, de whale tookee and carry him to Ninevah three days and three nights. When he got to de Ninevah he heave him on de shore. Ain' no shade in de seashore, so God suffer de gourd vine grow over he head for de shade.

Jonah wont go (to Ninevah) so God sendee de worm and cutee gourd vine (slashing gesture) down (hand lifted straight up). God said, "Jonah, your name called." He say to him, "Go in de Ninevah, and when he got dere he say forty days and forty nights and Ninevah shall be overthrowed."

And de king say, "Dis is de man of God—three days, three nights, de cow, de pigs neither de mules neither de chickens give em nothin to eat. Nobody eat neither drink."

So Jonah went to de mountain to see how it goin' be over throw, but stid uh dat God blessee dem. So den Jonah got mad: Say, "Lor', didn't you tell me you goin' 'stroy dat city?"

God say, "Jonah, dere's seven thousand women and chillun in dat city don't know right from wrong. If you think I go 'stroy dem, youse crazy."

How long Ninevah de blessee, I don't know. Dat de end right dere. Dats de fur I kin go.

Now Disa Abraham Fadda de Faitful

He had nephew name Lot—now dass right. Boffe of dem kinfolks. Dey have servant mind de stock whut dey raisee. One day dey two servants dey were quarreling.

Abraham say to de Lot, "We two kinfolks. Dese servant dey quarrel, don't lettee dat breakee de friendship. Now, data right, dasa left. Now which way you goin'?" Lot say to de Abraham, "I goin' to Sodom and Gomorrah, where you goin'?" Abraham say, "I goin' to de Land of Caanan."

When dey so much in sin in Sodom and Gomorrah, den de Lord he tookee two angels to pass Abraham's tent. Abraham seen dem and want to bow to dem and den he went and get kid and dressee him and set it before dem to eatee dinner. When dey get thew eatee dey start to Sodom and Gomorrah.

One de angels say to de udder, "Less not hidee our business from Abraham. Less tellee him whar we gwine." So dey say, "Abraham, do you know we goin' to Sodom and Gomorrah to settee it afire, goin' burn de place down? So muchee sin wentee before God dat God goin' burn de place out."

"Naw," Abraham say, "if I findee fifty ratcheous will you spare de city?" De angels say, "Yes, for your sake."

Abraham went to Sodom and Gomorrah and can't find de fifty ratcheous. "If I findee forty ratcheous will you spare de place?"

Dey say, "Yeah, for your sake, we spare 'em."

He fell back to twenty-five and couldn't find 'em. When he call for ten de Lord wont lissen. He flee way from him. Den de two angels go to Lot house and tell him, "Now you leavee here and don't lookee back."

When de people see de daughters of Zion come to Lot house, dey say to Lot, "Whut is dey doin dere?" Lot say don't bother dem. Den de angel pull Lot backee and wavee de hand and all de people go blind. Den dey say to Lot, "You flee away from here jes' as quick as you kin, and don't lookee back."

Lot's wife lookee back and turn to a pillar salt and she be dere till Judgment Day. Poor Cudjoe, I no lookee back. I pressee forward.

The Lion Woman

Three men, dey each have a lady. One say, "If I live to marry a wife, when she have a son, he go git down on top of a elephant to ride."

Another one say, "If I live to have a wife, when she have a son, he go git down on top a zebra for a ride."

De third man he say, "If I live to marry disa girl I love, when she have a son, he go git down on top a lion for a saddle horse."

De people, dey say, "How he goin't do dat? He cain't do dat because befo' he ketchy de lion, de lion ketchy *him*."

He say, "Oh, no!"

Well he marry de girl and dey have a son. When de boy he git so he kin run and throw de spear from the hand, you unnerstand me, de man he go in de woods and he found two young lions; but dey mama she gone killee something for them to eat. So he takee de two lions and killee one and takee de hide and stretch hit on de fence in de garden. De other one, you unnerstand me, he chain by de neck to de tree.

De mama lion she come home and she miss her babies, and she know de man take her children.

She feel hurtee, you unnerstand me, her breast swell way laka dis. She make up her mind she goin' punish de man whut killee her babies. So she turn herself into a woman, and many men see her come into de village. She look very fat and handsome and all de men want to marry her.

She tote a purse here (upon her hip). She say she will marry de man dat throw somethin' in de purse.

Everybody dey chunking at de purse. Dey chunk and dey chunk. Some throw too fur, some don't throw fur 'nough. Nobody make it go in de purse.

De man dat ketchee de lions, he stand and lookee but he don't try chunkee in de purse. He love his wife and don't want no mo' wife. She watch him and she aska him, "Why *you* no try chunkee in de purse? Don't you want me for yo' wife?"

He say, "I don't wanta chunk. I gotta wife already."

She say, "But I wanta you to chunk." She beg him please till after while jus' so he pick up somethin' wid his left hand throw disa way, but it went right in de purse, so she went home wid him to his house.

Soon's she git in de house she see de skin stretch on de garden fence and see de other one chained to de tree, and she swell up insider her, and she wish for night to come. She wishee it was night dat minute.

She lay in de bed wid de man dat night, but she ain' never go to sleep. He go sleep; but she wait to kill him. When she see he sleep, she turn back to a lion and got up walking in de house.

De man he got dogs, you unnerstand me, and dey know she a lion, and dey know when she git up to kill him. Jus' when she go to him to tear him up, de dogs bark and say, "No, you don't! No, you don't! Dass my master, and iffen you kill *him* you can't cross dis yard. We killee *you*."

She come back and lay down wid de man and wakee him up. She say, "Husband, I can't sleep. Yo' dogs makee so much noise, dey keep me wake. I think dey goin' come in de house and bite me. You betta go chain dem up."

He git up and go chain de dogs lak she say, den he go back to sleep. She git up agin, but de dogs hear her and dey talk so loud she skeered he hear 'em. So she git back in de bed and she think whut she kin do to kill him.

In de morning she say to him, "I can't stay wid you 'cause yo' dogs dey won't lemme sleep. I'm goin' home disa morning. You going piece de way wid me?"

He say he go wid her piece de way. He go git his

hunting spear and his bow and arrow, but she say, "Whut for you take de spear? You mean to killee me on de way? You don't need no arrow neither."

He tellee her he always take his spear when he go to de woods, but she cry and say she skeered he goin' kill her, so he put down de weapons. Den he put on his hunting knife but she make him take dat off, too. Den he takee a whistle, you unnerstand me, and put it in his shirt, and takee nine eggs to eat on de way. Den he go on wid her.

On de way dey talk. She aska him, "If a lion jump on you, whut you goin' do?"

He say, "I turn to a deer and run away fast."

"Oh, but a lion overtake a deer, den whut you do?"

"Den I turn to a snake and go in de ground."

"Oh, but de lion ketchee you befo' you dig de hole."

"Well, den I turn . . ." he start to say he turn to a bird and fly up in de tree, but de voice of his father come to him and say, "Hush!" so he say, "I don't know whut I do den."

After a while dey come to a woods and de woman excuse herself and go in de bush and stay a minute— den a big lion come out and take right after de man. He think quick whut he goin' do, and he turn to a bird and fly up in de highest tree.

De lion open one side and took out nine men wid

dey axes and open de other side and take out nine mo'
and dey 'gin to choppee down de tree. De man he blow
on de whistle so his dogs hear him and come.

De men dey chop hard at de tree. De lion she walk
round and round and roar whut she goin' do when de
tree fall. When de tree 'gin to fall, de man drop one egg
and de tree it come back up agin. He blow and blow for
his dog, but dey ain' heared him yit.

He drop another egg when de tree commence to fall
nex' time, and he kep' on till de last egg it gone. De
tree 'gin to shake agin, but he blow and blow on his
whistle.

One young dog say to de other, "Dat seem lak master's whistle I hear—don't you think so?"

De ole dog say, "Oh, lay down! You always hear
somethin' so you kin run in de woods."

After a while de young dog say he hear somethin'
agin, but de old dog say, "No, be quiet."

De tree is almost choppee down, and de lion stand
on her hind legs so she grab him when he fall. De
young dog say agin he hear de whistle and de ole dog
say, "Wait, I believe I hear somethin', too. Wait a minute." He lissen, den he say, "Hit *is* master's whistle! He
in trouble, too. Lemme go in de house and put de eye
medicine in de eye."

He go in de house and put de medicine in his eye, so

dat he kin see clear cross de world. "Unhunh!" he say. "I see master and he in bad trouble. Less go."

Dey run to de tree faster dan anythin' in de world and kill de lion and all de men. De man flew down from de tree and turn back to hisself agin. Den de man and de dogs take up all de meat and take it home and throw it in de yard. Den de man he go in de house wid his wife, but he don't tell her nothin' 'cause de ole dog he tell him dat if he tell, he will die.

When she look in de yard and see all de meat, she say to him, "Where you git all de meat?"

And he say, "I been hunting," but he don't tell her dat de dogs done made baskets outa plum twigs and brung de meat home. Dey walk on dey hind legs laka men and tote de baskets wid dey front legs.

His wife say, "You never brung home all data meat. No man kin tote so much, it too much for one man. You tell me who brung dat meat for you."

All day she keep dat up. Night time come and he wanta go to bed. She say no, she not sleep wid him never no mo' less he tell her 'bout de meat. So he tell her and den she sleep wid him. But de nex' mornin' she say to de dogs, "Why don't you tell me you kin tote meat laka man? Here I been had to wash yo' eatin' trough and tote yo' grub to you, and you plenty able to bring yo' plate and fetch yo' own grub."

Den de man he die 'cause he told whut de dog tell him not to, and de people make a great funeral for three days wid him. His wife she cry and cry 'cause she make him die, but dey go to bury him. But de ole dog say, "No, wait till his father come—he gone away on a journey." So dey wait three mo' days and when de father come he rub medicine on his eyes and he woke him, and he live a long time after dat, and his son git down on de lion he brung home.

Kazoola—Last of the slaves run thru the blocade from Africa—Landed in Mobile, Alabama, August 1859.

Cudjo Lewis (Oluale Kossola), in front of his home in Africatown (Plateau), Alabama, circa 1928. To have his photograph taken, Kossola dressed in his best suit and removed his shoes: "I want to look lak I in Affica, 'cause dat where I want to be."

Afterword
and Additional
Materials Edited
by Deborah G.
Plant

Afterword

H urston described Kossola as a "poetical old gen-
tleman . . . who could tell a good story."[1] And in
the tradition of the griot of his West African homeland,
Kossola tells a story of epic proportion. He is at once
storyteller and heroic figure, as he is the protagonist in
the saga he relates to Hurston. He was "left to tell" the
story of a massacre that befell the town of Bantè, and
he was the last original founder to sing the paean of Africatown. Characteristic of griots is their extraordinary
memory. As with others who had interviewed Kossola,
Hurston, too, took note of this attribute. In the preface to *Barracoon*, Hurston commends his "remarkable
memory." And she states, "If he is a little hazy as to
detail after sixty-seven years, he is certainly to be pardoned." Hurston used secondary sources in relation to

Kossola's narrative, but not as a corrective. Her use of historical research did not align with that of "the scientific crowd." "Woodson knew that people's memories were notoriously unsound and must be checked carefully by reference to written documents."[2] But Hurston's motivations were different: "The quotations from the works of travelers in Dahomey are set down, not to make this appear a thoroughly documented biography, but to emphasize his remarkable memory."[3]

"CUDJO'S OWN STORY"

Prior to their December 1927 meeting, Hurston had interviewed Kossola once before. As she states in her introduction to *Barracoon*, "I had met Cudjo Lewis for the first time in July 1927. I was sent by Dr. Franz Boas to get a firsthand report of the raid that had brought him to America and bondage, for Dr. Carter G. Woodson of the *Journal of Negro History*."[4] From February to August of 1927, Hurston conducted fieldwork in Florida and Alabama under the direction of Franz Boas, her mentor, the renowned "Father of American Anthropology." Boas had early on approached Woodson, the "Father of Black History," about a fellowship for Hurston, in support of the research. In accordance with their arrangements, Hurston was to collect black

folk materials for Boas and scout around for undis-
covered black folk artists. In addition to the gathering
of historical data for Woodson, she was also to collect
Kossola's story.[5]

Woodson supported Hurston's field research with a
$1,400 fellowship. Half of the funds came from the As-
sociation for the Study of Negro Life and History, an
organization founded and directed by Woodson. Elsie
Clews Parsons, of the American Folklore Society,
granted matching funds. As a fellow and "investigator"
for the association, Hurston was expected to contribute
material to the *Journal of Negro History*, a publication of
the association. During the latter part of her time in the
field, Hurston drove to Plateau, Alabama, to undertake
her last task for Woodson and conduct the interview
with Kossola. Along with various reports and archival
data, Hurston submitted to Woodson materials she had
collected on Fort Mosé, a black settlement in Saint Augustine, Florida. Woodson published this material as an
article entitled "Communications," in the October 1927
issue of the *Journal.*

In the same issue, he published Hurston's Kossola
interview as "Cudjo's Own Story of the Last African
Slaver."[6] A footnote at the beginning of the article
stated that as "an investigator of the Association for the
Study of Negro Life and History," Zora Neale Hurston

had traveled to Mobile to interview Lewis, "the only survivor of this last cargo." The note states further, "She made some use, too, of the *Voyage of Clotilde* and other records of the Mobile Historical Society."[7] In reality, Hurston made more than a little use of the society's records. And though part of the article was "a first-hand report," the larger portion of the article was secondhand information drawn from Emma Langdon Roche's *Historic Sketches of the South* (1914). Emma Roche was a writer, artist, and farmer born in Alabama in 1878. Her book is an account of the origins of slavery in America, couched in proslavery tenets and paternalistic perspectives. Her narrative recounts the history of the *Clotilda* and follows the fate of the Africans who were stored in its hold.

Only decades later would the literary critic and Hurston biographer Robert Hemenway bring the matter of Hurston's "borrowing" to scholarly attention and discussion. Hemenway credits the finding to the linguist William Stewart, who discerned it in 1972. "Stewart's discovery was conveyed to me," Hemenway noted in *Zora Neale Hurston: A Literary Biography*, "by John Swed of the University of Pennsylvania. I am grateful to Professor Stewart for granting me permission to cite his research and findings."[8] Though the footnote in her

1927 article acknowledges the Mobile Historical Society as a secondary source, it does not reference *Historic Sketches* specifically, and Hurston makes no direct reference to Roche's book within the body of the article itself. Rather, improperly documented paraphrased passages and near-verbatim appropriations from Roche's work constitute the larger part of the article. "Of the sixty-seven paragraphs in Hurston's essay," Hemenway relates, "only eighteen are exclusively her own prose."[9]

Hemenway speculates that Hurston found her interview with Kossola lacking in original material and therefore resorted to the use of Roche's work to supplement it. He supposes, too, that Hurston, writing at the outset of her career, suffered a quandary of purpose, direction, and methodology: How, exactly, was she to introduce the world to African American folklore, which she perceived to be "the greatest cultural wealth on the continent"?[10] Hemenway observed that Hurston, as one of the folk herself, struggled to negotiate the sociocultural chasm between her rural hometown of Eatonville, Florida, and the wealthy enclaves of New York City. He believed that her frustration with the academic study and presentation of the African American folk and folk culture was a reflection of the same struggle.

Hurston had imbibed Boas's theory of cultural relativity and understood that there were no superior or inferior cultures; she understood that cultures were to be assessed and evaluated on their own terms. But were the methods of Boas and Woodson conducive to her purposes? Was it possible that "the reportorial precision" of Western scientific investigation could be the means by which she would document and celebrate African American genius and, thereby, challenge European imperialism and Euro-American cultural hegemony? Or, did she believe, as did poet Audre Lorde, that *"the master's tools will never dismantle the master's house"*?[11]

In a letter to her friend Thomas Jones, the president of Fisk University, Hurston articulated her conundrum. "Returned to New York and began to re-write and arrange the material for Scientific publications, and while doing so, began to see the pity of all the flaming glory of being buried in scientific Journals."[12] She was dubious about Boas's objective-observer approach to folklore collection, and she chafed under Woodson's brand of scholarship. She preferred to be in the field, writes Hemenway, and so resented the time she spent investigating court records and "mindlessly transcribing historical documents."[13]

Nonetheless, Hemenway wondered why Hurston

would risk her career and whether her plagiaristic use of Roche's work was "an unconscious attempt at academic suicide." This attempt, Hemenway concludes, "is made because of a lack of respect for the writing one has to do." If detected and "her scientific integrity destroyed . . . Hurston's academic career would have been finished." She would then have been free from Boas's admonitions and Woodson's demands, and "the unglamorous labor" of collecting folklore.[14] Is it possible, Hemenway speculates further, that footnotes referencing Roche had been included but were lost or otherwise omitted from the "other records" to which the article's footnote alludes? In any case, Hemenway states that "Hurston's career needs no absurd apologetics. She never plagiarized again; she became a major folklore collector."[15]

Hurston biographer Valerie Boyd has proposed that even though Hurston resented the "hack work" she did for Woodson, it is also just as likely "that Hurston believed the report was only for Woodson's files; she did not expect it to be published any more than she thought her transcribed 'Communication' was worthy of publication."[16] The "Communications" article was a compilation of transcribed excerpts from letters and historical and congressional documents, strung together

with brief transitional statements. This style of report-
ing bears comparison with the composition of "Cudjo's
Own Story."

Boyd wondered whether Hurston's submission of
material that contained only 25 percent of her original work might have been Hurston's "way of getting
back at Woodson for arbitrarily slicing her pay and cutting into her research time by having her do his dreck
work."[17] Hurston had complained to her friend, the
poet Langston Hughes, that she had finished her work
for Woodson but wasn't paid in full. "I thought I'd get
pay for the month but he only paid me for two weeks."
She vented to Hughes and told him that she felt de-
pressed about the matter.[18]

As Hemenway conjectured that Hurston may have
saved the "juicy bits" of her folklore finds for theatri-
cal collaborations with Langston Hughes, Boyd conjectured that Hurston "had resolved to save her most
compelling material from Cudjo Lewis for her own
work."[19] Kossola had gained some celebrity as the last
living survivor of the *Clotilda*. Other anthropologists,
folklorists, historians, journalists, and artists alike
had sought him out. Hurston's colleague Arthur Huff
Fauset had already collected from Kossola the folktale
"T'appin" ("Terrapin"), which he published in Alain
Locke's 1925 *The New Negro: An Interpretation*.

Speculations aside, Boyd states, "Making 'some use' of material from another writer is completely common and acceptable. But, as Zora knew, copying another's work, and passing it off as one's own, is not."[20]

It is possible that the compromised article may have both relieved Hurston of tedium and allowed her a boon of lore for her own purposes, thereby hitting a straight lick with a crooked stick. Or, as Lynda Marion Hill suggests in *Social Rituals and the Verbal Art of Zora Neale Hurston*, Hurston's professional faux pas may have been an instance of Hurston masking her emotional response to a troubling event. In 1927, Zora Neale Hurston was new. Although Hemenway may have agreed with Franz Boas that "Hurston was a 'little too much impressed with her own accomplishments,'" it is equally true that she herself was still very much impressionable.[21] In 1927, the career to which critics allude was in the future. Hurston was not the seasoned social scientist who had published the folklore collections *Mules and Men* (1935) and *Tell My Horse* (1938). She was not the author of four novels, including the celebrated *Their Eyes Were Watching God* (1937). She was yet at the beginning of things.

"Cudjo's Own Story" was Hurston's debut scholarly publication. "In writing her first essay on Cudjo," Lynda Hill surmises, "Hurston might have been too moved and too uncertain how to manage her subjective response, rather than too frustrated with the rigors of scientific analysis, to produce an authentic text."[22] As Hurston reflected on her interview with Kossola years later in her autobiography, *Dust Tracks on a Road*, "It gave me something to feel about."[23] The interview changed Hurston, Hill observes. This elder, an Isha Yoruba in America, had schooled her in the sociopolitical and cultural complexities of "My People." In face of Kossola's recollections, the social constructions of "My People" and "Africans" were deconstructed by the reality of ethnic identifications, which not only distinguished tribes and clans but also generated the narrative distance and the ideological difference that rendered one ethnic group capable of regarding another as "stranger" or "enemy," and allowed that group to offer up the "Other" to "the trans-Atlantic trade."

"One thing impressed me strongly from this three months of association with Cudjo Lewis," Hurston writes. "The white people had held my people in slavery in America. They had bought us, it is true and exploited us. But the inescapable fact that stuck in my craw, was: my people had *sold* me and the white people had bought me. That did away with the folklore I had been brought up on—that the white people had gone to

Africa, waved a red handkerchief at the Africans and lured them aboard ship and sailed away."[24]

Hurston was a collector of folklore. However, the folklore she was "brought up on" contradicted the folklore she was collecting from Kossola. Moreover, "all that this Cudjo told me," Hurston mused, "was verified from other historical sources."[25] Harlem Renaissance pundits and artists like Zora Neale Hurston were wrestling with the identity of "the Negro." They had reclaimed the image of black people and asserted the value of black culture (vis-à-vis white people and Anglo American culture). There was a decided movement to do away with the image of "the Old Negro" and usher in "the New Negro," whose authentic culture and ethos were rooted in African origins. How did the butchering and killing of African "others" and the extirpation of whole societies fit within the profile of this modern, authentic "New Negro"?

Might Hurston have attempted to avoid "the inescapable fact" of that dimension of African humanity that was motivated by "the universal nature of greed and glory"?[26] Could it be that the woman and social scientist whose objectives entailed the discovering and uncovering of African cultural retentions in America was blindsided by Kossola's recollection of the inhumanity that was integral to his delivery at the port of Ouidah? Perhaps, rather than force herself to deal with such disorienting facts that stuck in her craw, Hurston chose, in the moment, to submit a narrative about the raid that had already been penned.

"Although justifying plagiarism is impossible," Hill writes, "the reasons for it should be scrutinized in light of its being, to date, a one-time occurrence in the long, productive career of a prolific and widely published author."[27] Hill's perspective is an important one, especially given the fact that Hemenway levels a similar charge, condemning and dismissing *Barracoon*, as though the manuscript were but an extension of the earlier published Cudjo Lewis piece. It is not. Hemenway proclaims that the article published in the *Journal* was an anomaly and reports that Hurston returned to Mobile to interview Kossola anew and did so with greater success. *Barracoon*, the book-length work, was the result of her efforts.

"Yet, even this unpublished manuscript, written in 1931," writes Hemenway, "makes extensive use of Roche and other anthropological sources; although it skillfully weaves together the scholarship and Hurston's own memories of Cudjo, it does not acknowledge those sources, and it is the type of book that Boas would have repudiated." Hemenway writes further, "The book purports to be solely the words of Cudjo; in fact, it is Hurston's imaginative recreation of his experience. Her purpose was to recreate slavery from a black perspective . . . but she was doing so as an artist rather than as a folklorist or historian."[28]

Although the journal article and the book manuscript have a common subject in Kossola, they are two distinct works. And where the charge of plagiarism is reasonable with the first, it is unfounded with the second. Hurston does draw on Roche's work in *Barracoon*, and she acknowledges it only indirectly. In her preface to *Barracoon*, she writes, "For historical data, I am indebted to the *Journal of Negro History*, and to the records of the Mobile Historical Society."[29] In her introduction, Hurston describes her interviews with Kossola and states, "Thus, from Cudjo and from the records of the Mobile Historical Society, I had the story of the last load of slaves brought into the United States."[30]

In her use of Roche's work, as with her use of other secondary materials, Hurston makes a good-faith effort in *Barracoon* to document her sources. She does *paraphrase* passages from *Historical Sketches*, and she places direct quotes within quotation marks, though in the manuscript draft she is inconsistent in this. And some sources are actually documented within the text of the introduction and others are footnoted within the body of the narrative.

The historian Sylviane Diouf states that Hemenway's characterization of Hurston's manuscript was "uncalled for." "She may have conflated some of what Cudjo said with some of what she knew as a scholar, but she made a genuine effort at separating the two. With few exceptions, the information provided in *Barracoon* is confirmed by other sources. Witnesses, experts in Yoruba cultures, contemporary newspaper articles, and abundant archival material corroborate the various events in Cudjo's life as described in *Barracoon*."[31]

Far from being a fictionalized re-creation, Diouf writes, "Cudjo's story, as transmitted by Hurston, is as close to veracity as can possibly be ascertained with the help of other records." She states further that Hurston "had produced an invaluable document on the lives of a group of people with a unique experience in American history."[32] Rather than repudiate her, Boas might well have been pleased and encouraging, as Hurston, in this early phase of her professional writing, endeavored to utilize historical records to support her folklore findings—just as both Boas and Woodson had instructed. What is more significant is that Hurston was struggling to appease neither Boas nor Woodson, but was engaged in the process of actualizing her vision of herself as a social scientist *and* an artist who was determined to present Kossola's story in as authentic a manner as possible.

HISTORIC DOCUMENT

From the earliest known "slave narrative" to the postbellum oral histories collected in works like George P. Rawick's *The American Slave*, one glimpses the vicissitudes and the interior lives of a people forced to exist in and toil under inhumane circumstances. Few of these narratives recount the incidents that preceded disembarkation and the holding pens and auction blocks of America. There are the journals of captains and manifests of ships, and there are the letters, diaries, bills of sale, and estate wills of the merchants and rulers of plantocracies who trafficked in African lives. As Hurston bemoaned in her introduction to *Barracoon*, "All these words from the seller, but not one word from the sold. The Kings and Captains whose words moved ships. But not one word from the cargo. The thoughts of the 'black ivory,' the 'coin of Africa,' had no market value. Africa's ambassadors to the New World have come and worked and died, and left their spoor, but no recorded thought."[33]

The subject of capture in Africa and transport through the Middle Passage is not the experience of those who were born into the condition of servitude on American soil. Narratives like Kossola's, of which there are but a few, describe the Maafa, the violent uprooting of bodies, the devastation of societies, and the desolation of souls. Rather than chart the journey from slavery to freedom in America, Kossola's narrative journeys back to Africa and gives us a glimpse into the collective black experience as seen through the openings in the barracoons that lined the African coasts of the Atlantic world.

Barracoon differs from classic slave narratives in a number of ways. The *Barracoon* narrative is not a conventional bid for freedom and it chronicles no harrowing tales of escape or trials of self-purchase. Unlike the authors of conventional narratives, Kossola was born in Africa. And because he was not born in the United States, he had to obtain citizenship through the naturalization process. Where narratives like those penned by Frederick Douglass speak to the cause of abolition, racial equality, and women's rights, *Barracoon* does not articulate an explicit political agenda. And it does not speak with the kind of heroic, self-possessed, and self-realized voice associated with black autobiography.

Where conventional slave narratives speak of conversions to Christianity, Kossola's narrative does also, but it does so while simultaneously expressing the spiritual traditions and customs of his homeland. He hadn't built up his hope on a future heavenly glory, but rather on a return to his people, a vision that speaks to the centrality of ancestral reverence. Kossola's nineteen years of life in Africa were more real to him than a declaration of independence in America. His narrative does not recount a journey forward into the American Dream. It is a kind of slave narrative in reverse, journeying backward to barracoons, betrayal, and barbarity. And then even further back, to a period of tranquility, a time of freedom, and a sense of belonging.

The African diaspora in the Americas represents the largest forced migration of a people in the history of the world. According to Paul Lovejoy, the estimated number of Africans caught in the dragnet of slavery between 1450 and 1900 was 12,817,000.[34] The Nobel laureate Toni Morrison dedicated her novel *Beloved* to "the 60 million and more," a number inclusive of the "disremembered and unaccounted for" in the Middle Passage.[35] Millions suffered capture and survived the passage across the Atlantic, but only a small number of Africans recounted their experiences.

As Sylviane Diouf points out, "Of the dozen deported Africans who left testimonies of their lives, only

[Olaudah] Equiano, [Mahommah Gardo] Baquaqua, and [Ottobah] Cugoano referred to the Middle Passage."[36] Eight of the ten narratives collected in Philip Curtin's *Africa Remembered: Narratives by West Africans From the Era of the Slave Trade* (1967) recount experiences of the Middle Passage. "They give us some notion of the feelings and attitudes of many millions whose feelings and attitudes are unrecorded," writes Curtin. "Imperfect as the sample may be, it is the only view we can recover of the slave trade as seen by the slaves themselves."[37] Ten years after Curtin's work, the scholar Terry Alford would exhume from the bowels of oblivion the events of the life of Abd al-Rahman Ibrahima, published as *Prince among Slaves: The True Story of an African Prince Sold into Slavery in the American South.* His narrative, too, recalls capture and deportation.

A few enslaved Africans like Olaudah Equiano, who experienced the Passage, acquired the skills to write their own narratives. Others like Kossola, who never learned to read or write, utilized the as-told-to mode of narration. Through this publication, *Barracoon* extends our knowledge of and understanding about the experiences of Africans prior to their disembarkation into the Americas. Like a relic pulled up from the bottom of the ocean floor, *Barracoon* speaks to us of survival and persistence. It recalls the disremembered and gives an account for the unaccounted. As an expression of the feelings and attitudes of one who survived the Middle Passage, it is rare in the annals of history.

THE MAAFA

"There is a loneliness that can be rocked," says the narrator in *Beloved*. "Arms crossed, knees drawn up; holding, holding on, this motion, unlike a ship's, smooths and contains the rocker. It's an inside kind—wrapped tight like skin. Then there is a loneliness that roams. No rocking can hold it down. It is alive, on its own. A dry and spreading thing that makes the sound of one's feet going seem to come from a far-off place."[38] It settles into the disjointedness of lives torn asunder by "a sequence of separations"; into the woundedness of a radical and "unbearable dislocation" from home and kin to an estranged place on foreign soil. The loneliness that attends such disruption infuses Kossola's narrative. It cannot be rocked. "After seventy-five years," writes Hurston, "he still had that tragic sense of loss. That yearning for blood and cultural ties. That sense of mutilation."[39] It is the existential angst that is consequent to deracination.

Maafa is a Ki-Swahili term that means disaster and

the human response to it.[40] The term refers to the disruption and uprooting of the lives of African peoples and the commercial exploitation of the African continent from the fifteenth century to the era of Western globalization in the twenty-first century. Conceptually, the phenomenon of the African Maafa is comprehensive in that it recognizes the extensive and continuous devastation of the African continent and its inhabitants and the continuous plundering that extends the trauma brought about via trans-Atlantic trafficking. For "illegitimate trade" was superseded by the European "scramble for Africa" and colonization of the continent, just as the "Peculiar Institution" of slavery in America was reformulated as the convict-leasing system, an earlier form of the Jail-Industrial Complex. And just as Kossola was ensnared in the institution of slavery in America, his son, Cudjo Lewis Jr., who was sentenced to five years of imprisonment for manslaughter, was handed over to the convict-leasing system in the state of Alabama.

Oluale Kossola could never fathom why he was in "de Americky soil." "Dey bring us 'way from our soil and workee us hard de five year and six months." And once free, he says, "we ain' got no country and we ain' got no lan'."[41] And in postbellum America he was subject to the exploitation of his labor and the vagaries of the law, just as he was in antebellum America. He remained confounded by this cruel treatment for the rest of his life. Kossola's experience was not anomalous. It is representative of the reality of African American people who have been grappling for a sense of sovereignty over their own bodies ever since slavery was institutionalized.

THE AMERICAN DREAM/
DREAMS DEFERRED

The American Dream is a major theme in the narrative of racial difference. The shadow side of that dream, which is not talked about, entails the plundering of racial "Others."

It was this dreaming that inspired both William Foster and Tim Meaher to flout the law of the US Constitution, steal 110 Africans from their homes, and smuggle them up the Mobile River and into bondage. Though Foster and Meaher were charged with piracy, neither was convicted of any crime. No one was held responsible for the theft of Kossola and his companions and their exploitation in America. Of the thousands of Africans smuggled into America after 1808, only one man was held accountable and hanged, and even he died proclaiming his innocence.

Folklore had it that Tim Meaher decided to smuggle

Africans into Alabama on a bet. In April of 1858, while traveling aboard the *Roger B. Taney*, Meaher boasted to fellow passengers that he could bring Africans into the country in spite of the ban against trans-Atlantic trafficking. He bet "any amount of money that he would 'import a cargo in less than two years, and no one be hanged for it.'"[42] It was Meaher's dream to own land and become wealthy and to use slave labor to do it. He believed it was his birthright.

AFRICATOWN

At the end of the Civil War, once they learned they were free, Kossola and his compatriots began to plan their repatriation. They soon realized that their meager earnings would not be adequate to live on and allow them to save enough money to fulfill their dreams of returning to Africa. Also unaware of the activities of the American Colonization Society, they resolved to re-create Africa in America. Toward that end, the community of Africans elected Kossola to approach Timothy Meaher about granting them some land on which to rebuild their lives as a free people.

"You made us slave," Kossola told Meaher. "Now dey make us free but we ain' got no country and we ain' got no lan'. Why doan you give us piece dis land so we kin buildee ourself a home?"[43] Meaher's response was one of indignation. "Fool do you think I goin' give you property on top of property? I tookee good keer my slaves in slavery and derefo' I doan owe dem nothing? You doan belong to me now, why must I give you my lan'?"[44] Kossola and the others rented the land until they were able to buy it from the Meahers and other landowners. The parcels they bought became Africatown, which was established by 1866.

Their African Dream was inextricably bound up with Timothy Meaher's American Dream, and their dream of return would be forever deferred. But the survivors of the *Clotilda* would work together to create a community that embodied the ethos and traditions of their homeland. In its founding and government, Africatown was similar to other black towns, writes Sylviane Diouf. But it was distinguished by the fact of its ethnicity. Although some African Americans were numbered among them as spouses and founders, Africatown "was not conceived of as a settlement for "'blacks,' but for Africans."[45]

Africatown was their statement about who they were, and it was a haven from white supremacy and the ostracism of black Americans. The bonds the Africans created in the barracoons, on the ships, and in servitude were the source of their survival and resilience, and the foundation of their community.[46]

Africatown is more than a historic site. It is a place expressive of African ingenuity and a prime model of the processes of African acculturation in the American South.

As Africatown is more than a cultural legacy, Oluale Kossola was not just a repository of black genius, tapped for a few stories, tales, and colorful phrases, and Zora Neale Hurston knew this. She did not perceive *Barracoon* as another cultural artifact illustrating the theoretical characteristics of Negro expression but as one, singular, portrait of black humanity. "Slavery is not an indefinable mass of flesh," as Ta-Nehisi Coates writes.[47] It is a particular and specific woman or man. It is Kossola, and his wife, Abilé, their six children, the host of Africans who founded Africatown, and their shipmates who survived the *Clotilda*.

We must courageously embrace this history because it is, as James Baldwin understood, "literally *present* in all that we do," and the power of this history, when we are unconscious of it, is tyrannical.[48] The history of Kossola's life elucidated, for Zora Neale Hurston, "the universal nature of greed and glory" as an "inescapable fact" of our common humanity. It is this common humanity that Hurston struggled to make the world understand.

If we view *Barracoon* as just another brilliant example of Hurston's anthropological genius, we are gravely mistaken and we do not fathom the full import of her objectives as a social scientist. In her endeavor to collect, preserve, and celebrate black folk genius, she was realizing her dream of presenting to the world "the greatest cultural wealth on the continent," while simultaneously contradicting social Darwinism, scientific racism, and the American pseudoscience of eugenics. She was refuting the tenets of biological determinism that were at the heart of the Great Race theory. The body of lore Hurston gathered was an argument against such notions of cultural inferiority and white supremacy, and it defied the idea of European cultural hegemony as it also questioned the narrative of white nationalism.

Barracoon is a counternarrative that invites us to break our collective silence about slaves and slavery, about slaveholders and the American Dream. Completed in 1931, the narrative of Oluale Kossola has finally found its audience, and Zora Neale Hurston's first book-length work has found a taker and is now finally published. Though nearly a century has passed between the completion of the final draft of her manuscript and the publication of *Barracoon*, the questions it raises about slavery and freedom, greed and glory, personal sovereignty and our common humanity are as important today as they were during Kossola's lifetime.

Acknowledgments

FROM THE ZORA NEALE HURSTON TRUST

The trustees of the Zora Neale Hurston Trust wish to thank those who contributed to the publication of Zora Neale Hurston's never-published work, *Barracoon*. We have no claim as the authors of this work; however, we are the custodians of Zora Neale Hurston's legacy, and, as such, we are committed to preserving her standing in the world as a literary icon and an anthropological giant. We gratefully acknowledge our agent, editors, and publishers, as well as the academics and devotees whose shared love of Zora's *Barracoon* led them to embrace publication of this work.

We are thankful for the efforts of the Joy Harris Literary Agency staff and for those of Joy Harris, our agent, who worked tirelessly to promote this work. Joy provided us with the guidance and the steady hand we needed to supply her with a publishable manuscript. Despite our sometimes moving in different directions, she was able to corral our activities so we could deliver the story. Joy loved Cudjo Lewis from the start and shared our faith that Cudjo's story was meant to be published. We also acknowledge Adam Reed, Joy's valued associate. He was a force in our effort to prepare a completed manuscript worthy of review by Joy and publishers. No job was too small for his attention.

For their recognition of *Barracoon* as an invaluable contribution to the story of slavery in America, we want to express our gratitude to our publishers at HarperCollins: Tracy Sherrod, the editorial director of Amistad; Jonathan Burnham, the publisher of Harper-Collins; and Amy Baker, the associate publisher of Harper Perennial and Harper Paperbacks. All of them determined that Cudjo Lewis's story had to be told, and they helped make it possible for *Barracoon* to be born. Additionally, we extend our appreciation to Diane Burrowes, senior director of academic and library markets, and to Virginia Stanley, the director of academic and library markets, who contributed their expertise to this publication.

To Deborah G. Plant, PhD, we extend our most

heartfelt appreciation for her editorial work. Deborah brought her love of all things Zora to this project. We are grateful for her diligence in researching issues related to the manuscript and for providing answers to questions that could be posed. We are also grateful for Deborah's appreciation and explanation of Zora's use of ethnographic methodology in telling Cudjo's story. She was in tune with Zora's energy throughout.

We extend our continuing gratitude to the many scholars who have been champions of Zora Neale Hurston. Without their love and advocacy, Zora's works and her personal vitality may have been lost to generations. We are grateful to Alice Walker, who became a crusader for Zora and pronounced her "a Genius of the South." We are grateful to Cheryl Wall, who knows so much about Zora and has generously shared her findings with others. We are grateful to Valerie Boyd, who helped us know, understand, and love Zora through her biography of Zora's life. We are grateful to Kristy Andersen, who introduced so many to Zora through her documentary work on Zora's life.

We owe a debt of gratitude to Howard University's Moorland-Spingarn Research Center and its curator, Joellen ElBashir, for serving as the custodian of the manuscript of *Barracoon* for so many years. We are also grateful to the Mobile Historical Society for providing historical documents that certify the life of Cudjo Lewis in America.

We can never repay those who have loved and supported Zora in her quest to leave us with a cultural legacy on many levels, but we can rejoice with them in celebrating Zora's acceptance today as one of the world's foremost folklorists as well as a literary genius. *Barracoon* is a perfect example of Zora's talent in many genres. It is a late publication, but it is timely in its instruction.

FROM DEBORAH G. PLANT

I am forever appreciative of the legacy of Zora Neale Hurston and grateful for her magnanimous spirit. I am thankful for the direction of Dr. Linda Ray Pratt, of the University of Nebraska–Lincoln, who was there at the beginning of things with me and my investigations into the life and works of Zora Neale Hurston. It is with an abundance of gratitude that I thank the members of the Zora Neale Hurston Trust (Lois Gaston, Lucy Ann Hurston, and Nicole Green) for the opportunity to be of service in the publication of Hurston's narrative.

For her support and assistance, I thank my sister, Gloria Jean Plant Gilbert, who traveled with me to Africatown and captured some of the spirit of the place in the photos she took. For their expertise, direction, patience, and kindness, I offer thanks and deep appreciation to Joy Harris of the Joy Harris Literary Agency and her associate, Adam Reed; and to HarperCollins editorial director Tracy Sherrod and her assistant, Amber Oliver.

I wish to thank those writers whose works have contributed to our knowledge about the Africans who were smuggled into the United States aboard the *Clotilda*, enslaved in Alabama, and who, in freedom, founded a town and left a rich heritage. I appreciate the generous spirit of Ms. Mary Ellis McClean, Kossola's great-granddaughter who spoke with us at the Union Missionary Baptist Church (organized originally as the Old Landmark Baptist Church in 1872), of which Kossola was a founding member. I offer especial thanks to Sylviane A. Diouf and Natalie S. Robertson for their groundbreaking research and publications on "the *Clotilda* Africans"; and to Lynda Marion Hill for her perceptive analysis of Hurston's efforts in writing the *Barracoon* narrative. I extend immense thanks and gratitude to Ms. Patrice Thybulle, my dedicated research assistant and collaborator on the Maafa Project, initiated during my tenure at the University of South Florida. I thank Howard University librarian and curator Joellen ElBashir for her assistance. And I express appreciation to my parents, Alfred and Elouise Porter Plant, and Roseann and Henry Carter for their inspiration; and I thank Phyllis McEwen, Gwendolyn Lucy Bailey Evans, Joanne Braxton, Virginia Lynn Moylan, Valerie Boyd, Cathy Daniels, Marvin Hobson, Lois Plaag, and Sam Rosales for their ongoing friendship and support.

I honor the ancestral spirit of Oluale Kossola (Cudjo Lewis) and thank him for his poignant life story.

Founders and Original Residents of Africatown*

"AFRICAN" NAME	AMERICAN NAME	ORIGIN
Pollee/Kupollee	Allen, Pollee (Pollyon)	Yoruba
	Allen, Lucy	
	Allen, Rosalie (Rose)	
Monabee (Omolabi)	Cooper, Katie (Kattie)	Yoruba
	Dennison, James	South Carolina

---

* This table is drawn from the works of Sylviane A. Diouf, *Dreams of Africa in Alabama: The Slave Ship "Clotilda" and the Story of the Last Africans Brought to America* (New York: Oxford University Press, 2007); and Natalie S. Robertson, *The Slave Ship Clotilda, and the Making of AfricaTown, USA: Spirit of Our Ancestors* (Westport, CT: Praeger, 2008).

"AFRICAN" NAME	AMERICAN NAME	ORIGIN
Kanko (Kêhounco)	Dennison, Lottie	Yoruba
	Dozier, Clara	
	Ely, Horace	Alabama
	Ely, Matilda	Alabama
	Johnson, Samuel	
	Keeby, Anna (Annie)	
	Johnson, Samuel	
	Keeby, Anna (Annie)	
Keeby, Ossa		Hausa
Gumpa	Lee, (African) Peter	Fon
	Lee, Josephine	
	Lewis, America (Maggie)	
Abila (Abilé)	Lewis, Celia (Celie)	Yoruba
Oluale (Oloualay)	Lewis, Charles (Char-Lee)	Yoruba
Kossola (Kazoola)	Lewis, Cudjo	Yoruba
	Livingston, John	
Ar-Zuma	Livingston (Levinson), Zuma	Nupe
	Nichol, Lillie	Africa/?
	Nichol, Maxwell	Alabama
Jaba (Jabi or Jabar)	Shade, Jaybee (Jaba)	Jaba/Jabi?
	Shade, Polly (Ellen)	
	Thomas, Anthony (Toney)	
	Thomas, Ellen	Alabama

"AFRICAN" NAME	AMERICAN NAME	ORIGIN
Abache (Abackey)	Turner, Clara Turner, Samuel Wigfall (Wigerfall), Hales	Yoruba
Shamba	Wigfall (Wigerfall), Shamba Wilson, Lucy	Shamba?

Glossary

Clotilda, **The**: A 120 81/91-ton schooner built by William Foster in Mobile, Alabama, in 1855. It was 86 feet long, 23 feet wide, and $6^{11}/_{12}$ feet deep. Two-masted, with one deck, it was built for speed. These types of ships were designed during the years of suppression of the traffic, in order to outmaneuver those ships that were policing the waters. The US Constitution declared those engaged in the illegal importation of Africans into America to be pirates and declared that those apprehended would be charged with piracy—and hanged. In collaboration with Timothy Meaher, William Foster refitted the *Clotilda* as a "slaver." Its journey to Africa represented their first smuggling venture, and it would be their last. In March 1860, Foster set sail for Ouidah on the coast of

West Africa, where he illegally bought 125 Africans who were held in the barracoons of Dahomey. Fearful of being captured by two approaching steamers, Foster weighed anchor and left fifteen Africans on the beach. After about forty-five days on the Atlantic, Foster docked near Twelve-Mile Island off the Mobile River. After disembarkation of the Africans, Foster burned and scuttled the *Clotilda* at Big Bayou Canot, in an effort to cover up his piracy. The *Wanderer,* which transported more than four hundred Congolese captives to Jekyll Island, Georgia, in November 1858, had long been considered the last vessel to import Africans illegally into the United States. With its documented 1860 arrival into Mobile Bay, the *Clotilda* now holds that unfortunate distinction.

Illegitimate Trade: A series of constitutional acts transformed trans-Atlantic trafficking from a "legitimate" to an "illegitimate" activity. American participation in trans-Atlantic trafficking can be traced to the colonial era. As the largest trafficking enterprises in the colonies at that time were run out of Rhode Island, the D'Wolf family, headed by James and Charles D'Wolf, ran the largest trafficking enterprise in Bristol, Rhode Island, after the American Revolution.

By the end of the eighteenth century, American vessels, along with the British and the Portuguese, would dominate the Atlantic traffic in human beings. In 1794, the US Congress passed legislation that outlawed the building of or fitting out of ships for the purpose of importing Africans into America or for trafficking enterprises in other countries. Penalties ranged from fines of $200 to $2,000. The March 1807 Act Prohibiting the Importation of Slaves declared all participation in international trafficking to be illegal and abolished the importation of Africans into the United States. Fines for violation were increased to upwards of $20,000 and imprisonment of at least five but not more than ten years. The act was to take effect on January 1, 1808. The 1820 act charged participants in the traffic with piracy, which carried a penalty of death. Although international trafficking had been deemed illegal or "illegitimate," proslavery adherents continued to engage in it. The United Kingdom also abolished trans-Atlantic trafficking in 1807. In its efforts to suppress the traffic in humans, it encouraged and promoted "legitimate trade" with Africa. Such trade entailed the exchange of "legitimate" commodities from Africa, such as the agricultural exports of palm oil, palm kernels, kola nuts, and ground nuts.

Jim Crow: *Jim Crow* refers to the social system that developed in the United States following the Civil War. The name "Jim Crow" is based on a character developed by "the Father of American minstrelsy," Thomas Rice, who performed in blackface. Rice appropriated the song about Jim Crow from black folklore and created a stereotypical character of blacks as lazy, ridiculous, worthless subhumans. Rice's derogatory depictions of black people were popular with his white audiences. The name "Jim Crow" then became synonymous with the system of racial segregation that cast blacks as inferior beings while elevating whites as superior. In 1896, the US Supreme Court's decision in *Plessy v. Ferguson* would sanction Jim Crow. The decision upheld the doctrine of separate but equal, which segregated the races in public spheres and effectively ushered in de jure segregation in American society.

Krooboys: Krooboys and Kroomen were a group of seafarers and ship laborers who settled along the West African coastline. They originate from the Kru (or Kroo) peoples of the Liberian hinterlands who migrated to the west coast. During the eighteenth century, they worked as sailors and laborers for the British and Europeans in their maritime commerce with West Africa. They worked aboard trafficking vessels and operated as dealers, brokers, and middlemen for those looking to purchase Africans. They were known for their skills in maneuvering canoes filled with people or merchandise through the rough surf, onto the beach, or out to ships.

Maafa: Marimba Ani defines *Maafa* as a Ki-Swahili term that means disaster and the human response to it. The term refers to the disruption and uprooting of the lives of African peoples and the continuous commercial exploitation of the African continent—from the fifteenth century to the era of Western globalization. The African Maafa entails the multidirectional, violent, and catastrophic phenomenon that pervaded the entire African continent, not just its western coast. Thus the concept also encompasses the trafficking of Africans across the Sahara, the Mediterranean Sea, the Red Sea, and the Indian Ocean, which occurred centuries before the commencement of trans-Atlantic trafficking.

Middle Passage: The Middle Passage describes the transoceanic route taken by trafficking vessels from the west coast of Africa to the Americas. It also refers to the middle leg of what is called "the triangular slave trade": Ships originating in England or Europe would sail to the African coast to exchange manufactured goods for African captives; Africans were then sold or exchanged in the Americas for raw materials (cotton, sugar, coffee, tobacco); ships laden with these materials would then make the return journey from the Americas to Europe. The length of the voyage from African shores to ports in the Caribbean and the Americas varied. The voyage from Africa to Brazil would take at least a month. From Africa to the Caribbean or North America could take two or three months. Other variables such as wind, inclement weather, mutiny, rebellion, or escape from other vessels would hasten or retard a ship's passage.

Mosé, Fort: In the late 1600s, Africans escaping enslavement in the British colonies settled in Spanish territory near Saint Augustine, Florida. In 1738, the Spanish governor, Manuel de Montiano, fortified the settlement with the construction of Fort Gracia Real de Santa Teresa de Mosé, granting the settlers citizenship and sanctuary and thereby establishing the first free black town in North America. The fort would become the northernmost point of Spanish defense against the British, and the townsmen would become members of the Fort Mosé militia. Captain Francisco

Menendez, who had escaped enslavement in South Carolina, was appointed as the "chief" of the town. Under his leadership, the Fort Mosé militia, along with Native Americans and the European residents of Saint Augustine, defended the fort against a British attack in 1740. Fort Mosé remained a haven for Africans, African Americans, and Native Americans until the Peace of Paris agreement of 1763 that ceded Florida to the British.

Orìṣà: In the spiritual traditions of the Yoruba people of West Africa, the supreme deity is manifested as the trinity of Olodumaré, Olofi, and Olorun. The Orìṣà are a reflection of these divine expressions. They represent a pantheon of deities who embody specific qualities of the cosmos. Among the pantheon are Obatalá, Oshún, Yemayá, Changó, Oyá, and Ogún. Traditional ceremonies serve to unite humans with the spirit realm and restore balance between humans and nature. Ancestral reverence is an integral aspect of the tradition. In the Americas, African spirituality was a source of resilience and resistance to the bleak and absurd reality into which African peoples had been forced. The Orìṣà tradition, along with other spiritual traditions of West African peoples, merged with the religious traditions of European Christianity and those of indigenous Amerindians to create new belief systems such as Vodun, Hoodoo, Obeah, Santería, and Candomblé. Zora Neale Hurston investigated and documented these syncretic religions in *Mules and Men* (1935) and *Tell My Horse: Voodoo and Life in Haiti and Jamaica* (1938).

Roche, Emma Langdon: Emma Roche was born on March 26, 1878, in Mobile, Alabama. She was the daughter of Thomas T. and Annie Laura (James) Roche. Emma Roche was an artist, writer, housekeeper, and farmer. She wrote *Historic Sketches of the South* in 1914 and illustrated the book with her own drawings and photographs of the residents of Africatown.

"Slaver": Trafficking vessels were called "slavers," because those involved in the trafficking did not see the Africans they transported as human beings, but instead as *slaves* (i.e., chattel, commodities, cargo, merchandise). And they treated them accordingly. Aboard these vessels Africans experienced shock or melancholia, knowing neither their destiny nor their fate. The holds were dark and fetid. In the beginning of these oceanic voyages, the mortality rate for Africans could be as high as 50 percent. During this period, "tight packing" was a common method of loading Africans into the ships. In order to minimize loss due to high mortality rates, captains had their crew cram as many people as possible into a hold, allowing little room to turn or sit up. On some ships, Africans were laid one on top of the other, stacked like logs. In later centuries, changes in the design of the ships, regulations, and the desire for greater profit would modify the methods used to transport Africans.

"Slavery": The term *slave* originally meant *captive*, and it was historically associated with the Slavic peoples of Eastern Europe, who were conquered by western Europeans in the ninth century and forced into conditions of servitude. The same term has been used in reference to African peoples whom western Europeans pressed into servitude in the Caribbean and the Americas. It has also been used to refer to the condition of servitude practiced in Africa prior to Arab-Islamic and European encroachments into Africa.

Characteristic features of slavery are that people perceived to be different from the larger society can be subjugated and exploited for their labor; that these people have no rights and are considered property, a thing owned; and that they and their offspring inherit this condition for life.

In the United States, slavery has been called the Peculiar Institution. As it was elsewhere in the Americas, this institution was violent, inhumane, and racialized.

"Slavery" (African "Internal Slavery"): Forms of servitude existed in Africa prior to the invasions of Arab Muslims and Europeans—but it was not slavery. Slavery was but one form of servitude or labor practiced in various civilizations from antiquity to the modern day. Serfdom, clientage, wage-labor, pawnship, and communal work represent other kinds and conditions of labor that were practiced. The conditions of labor in ancient or early African society were more characteristic of conditions associated with feudalism, not slavery, and were more aptly described as relationships of dependency.

Africans in conditions of servitude could be subject to labor others refused to perform, labor that was considered degrading, tedious, or dangerous. They could be subjected to maltreatment and even be used as living sacrifices. But for the most part, Africans in relationships of dependency had rights and maintained their human dignity. After the mid-fifteenth century, the systems of servitude among West Africans were transformed, as trans-Atlantic trafficking became integral to the politics and economy of African societies.

Trans-Atlantic trafficking also transformed the identity of people on the African continent and their relationship to one another. As people were now perceived as slaves, those outside a particular group—in terms of ethnicity, ideology, or lineage—became subject to capture and deportation. In spite of the ethnic and cultural diversity of the people on the continent, Europeans and Americans referred to them, collectively, as "Africans." This resulted in the belief that "'Africans' sold their own sisters and brothers." This tendency to generalize the varied ethnic groups as "Africans" has been a continuous source of conflict for the people of Africa and the African diaspora.

Trans-Atlantic Slave Trade: The business of ferrying captive Africans across the Atlantic to other lands for the cultivation of cash crops was initiated by the Portuguese in the mid-fifteenth century. Prince Henrique of Portugal (1394–1460), known throughout Europe as "Henry the Navigator," knew well of the vast wealth of Africa and Asia. No longer content with negotiating with Moors, Berbers, and Arab middlemen for the goods and people of sub-Saharan Africa, he sought direct access to these continents—not over land, but by sea. Before trade for African bodies was established and regularized, European seafarers engaged in the typical "smash and grab" approach of acquiring Africans for use as slaves. In 1441, the Portuguese seized twelve Africans from the west coast of Africa. Subsequent actions of that sort resulted in retaliation. The Portuguese then established formal compacts with African officials. Portuguese colonists, settled in the island of Madeira, had begun experimentation with the cultivation of sugarcane. Initially, they imported eastern Europeans and Africans to perform this labor. However, Constantinople's fall to the Turks in 1453 closed the "slave ports" of the Black Sea to western Europeans seeking eastern European slaves. In the aftermath of this turn of events, the majority of the laborers in the cane fields of Madeira came from the African continent. The Portuguese replicated this model of cultivating sugarcane with the labor of Africans in the plantations of the Caribbean and the Americas. Other European nations, England, and the English colonies in North America emulated the Portuguese. Well after the abolition of trans-Atlantic trafficking by most European nations and the United States, the Portuguese would persist in their trafficking enterprises until 1870.

Notes

INTRODUCTION BY DEBORAH G. PLANT

1. Zora Neale Hurston to Langston Hughes, 9 December 1927, in *Zora Neale Hurston: A Life in Letters*, ed. Carla Kaplan (New York: Doubleday, 2002), 110.

2. See Hurston's introduction, in the present volume. Hurston utilized different spellings of Kossola's name. In *Dust Tracks*, she used the syllabic spelling that was characteristic of her technique when recording dialect: "There I went to talk to Cudjo Lewis. That is the American version of his name. His African name was Kossola-O-Lo-Loo-Ay" (198). In the "Barracoon" manuscript, she consistently refers to him as "Kossula."

3. Lynda Marion Hill, *Social Rituals and the Verbal Art of Zora Neale Hurston* (Washington, DC: Howard University Press, 1996), 68.

4. Sylviane Diouf, *Dreams of Africa in Alabama: The Slave Ship "Clotilda" and the Story of the Last Africans Brought to America* (New York: Oxford University Press, 2007), 40. Biographical background of Kossola Oluale and the historical backdrop of the *Clotilda* is drawn from the works of Diouf; Natalie Robertson, *The Slave Ship "Clotilda" and the Making of AfricaTown, USA: Spirit of Our Ancestors* (Westport, CT: Praeger, 2008); and Zora Neale Hurston, "Barracoon: The Story of the Last 'Black Cargo,'" box 164-186, file 1, unpublished typescripts and handwritten draft, 1931, Alain Locke Collection, Manuscript Department, Moorland-Spingarn Research Center, Howard University.

5. It seems that "Oluale," Kossola's father's name, thus Kossola's African first name, became contracted into Lewis, which became Kossola's American last name. "Cudjo" is the name given a male child born on a Monday. This became Kossola's American first name.

6. Paul E. Lovejoy, *Transformations in Slavery: A History of Slavery in Africa*, 3rd ed. (New York: Cambridge University Press, 2012), 135–36; Robertson, *Slave Ship "Clotilda,"* 36–37.

7. Diouf, *Dreams of Africa in Alabama*, 30, 31.

8. Lovejoy, *Transformations in Slavery*, 141.

9. Robertson, *Slave Ship "Clotilda,"* 84.

10. Zora Neale Hurston, *Dust Tracks on a Road: An Autobiography* (Urbana: University of Illinois Press, [1942] 1984), 204.

11. Ibid., 202.

12. Hurston to Charlotte Osgood Mason, 25 March 1931, in Kaplan, *Letters*, 214.

13. Hill, *Social Rituals*, 72; Boyd, *Wrapped in Rainbows*, 167.

14. Cudjo Lewis to Charlotte Mason, Alain Locke Collection, Manuscript Department, Moorland-Spingarn Research Center, Howard University.

15. Hurston to Mason, 26 May 1932, in Kaplan, *Letters*, 257.

16. Hurston to Mason, 12 January 1931, in Kaplan, *Letters*, 201; Hurston to Mason, 18 April 1931, 217.

17. Hurston to Mason, 25 September 1931, in Kaplan, *Letters*, 228.

18. Langston Hughes, *The Big Sea: An Autobiography* (New York: Thunder's Mouth Press, [1940] 1986), 334; Boyd, *Wrapped in Rainbows*, 221.

19. Diouf, *Dreams of Africa in Alabama*, 3.

20. Hill, *Social Rituals*, 64.

21. Zora Neale Hurston to Langston Hughes, 12 April 1928, in Kaplan, *Letters*, 116.

22. Hill, *Social Rituals*, 65, 66.

23. Ibid., 65.

24. Ibid., 67.

INTRODUCTION BY ZORA NEALE HURSTON

1. Emma Langdon Roche, *Historic Sketches of the South* (New York: Knickerbocker Press, Scholar Select Reproduction, [1914] 2016), 72.

2. According to Diouf and Robertson, the boat was built and owned by William Foster.

3. Roche, *Historic Sketches*, 85.

4. Ibid., 86.

5. Ibid.

6. Ibid.

7. Ibid. Diouf and Robertson document that the majority of the captives taken were of Yoruba ethnicity. Among the other ethnic groups that made up the "*Clotilda* Africans" of Africatown, there was only one who was Fon, and that was Gumpa (African Peter). It was not the policy of the king of Dahomey to enslave subjects of his own kingdom. Giving Gumpa to Foster seems to have been an imperious lark.

8. Roche, *Historic Sketches*, 73.

9. Ibid.

10. Frederick Edwyn Forbes, *Dahomey and the Da-homans: Being the Journals of Two Missions to the King of Dahomey, and Residence at His Capital, in the Years 1849 and 1850, Volume 1* (Charleston, SC: Bib-lioBazaar Reproduction Series, [1851] 2008), 14, 15.

11. Ibid., 15, vi.

12. Ibid., 15–16.

13. Ibid., 17.

14. Ibid., 17–18.

15. [Editor's note: The scenario of Foster's manner of coming ashore parallels the description of a similar scene in Forbes's *Dahomey and the Dahomans*, which notes the difficulty of both maneuvering through the surf to shore and from the shore to a ship. In relating his experience, Forbes alludes to "three kroomen" and a "kroo canoe" that was "dashed to pieces." This particular scene likely proved significant to Hurston, as Forbes was meeting with a Mr. Duncan on a mission to persuade the King of Dahomey (Ghezo) "to consent to a treaty for the effectual suppression of the slave trade within his dominions" (45, 44).]

16. [Editor's note: Foster's narrative states that "while getting underway two more boats came along side with thirty-five more negroes, making in all one hundred and ten; left fifteen on the beach having to leave in haste" (William Foster, "Last Slaver from U.S. to Africa, A.D. 1860," Mobile Public Library, Local History and Genealogy, 9).]

17. Roche, *Historic Sketches*, 88. [Editor's note: According to Foster's account: "We had an alarming surprise when man aloft with glass sang out '*sail ho*' steamer to leeward ten miles" (8).]

18. Roche, *Historic Sketches*, 88.

19. Ibid., 89–90.

20. Ibid., 90–91.

21. Ibid., 94–95.

22. Ibid., 95, 96.

23. Ibid., 96, 96 n1.

24. Ibid., 96–97.

25. Ibid., 97. [Editor's note: "Bayou Corne" is a colloquialism referencing Big Bayou Canot.]

26. Ibid., 98–99.

27. Ibid., 99–100.

28. See Hurston's chapter 6 in the present volume.

29. [Editor's note: Although they were charged and fined, neither Meaher nor Foster paid any fine.]

CHAPTER I

1. [Editor's note: In *Dust Tracks on a Road*, Hurston writes that she "went to talk to Cudjo Lewis. That is the American version of his name. His African name

was Kossola-O-Lo-Loo-Ay" (198). Hurston also transcribed Kossola's name as "Kossula" and "Kazoola." In my introduction and references elsewhere, I have used "Kossola," as it is consistent with Sylviane Diouf's research in *Dreams of Africa in Alabama* where she establishes "Kossola" as the likely spelling: it is a name "immediately decipherable" to the Isha Yoruba who "have a town named Kossola" (Diouf, 40).]

CHAPTER IV

1. [Editor's note: Based on Kossola's description of the Dahomian raid and her research in Richard Burton's *A Mission to Gelele, King of Dahome*, vol. 1 (New York: Frederick A. Praeger, [1894] 1966), Hurston was convinced that Takon was Kossola's hometown, and therefore, the king in that town, Akia'on, must have been the name of Kossola's king.

Hurston understood the ethnic identity of Kossola and his compatriots to be Takkoi, a variation of "Tarkar." In *Historic Sketches*, Roche recorded the ethnicity of Africatown founders as "Tarkar." However, Diouf relates in *Dreams of Africa in Alabama* that no such ethnicity existed. Roche's book about the Africatown enclave had become "an obligatory reference for journalists and others. . . . Since she understood that they were 'Tarkars,' this pseudo-ethnicity has been repeated by reporters, scholars, and even the Africans' descendants" (246).

Believing the name of the Africans' ethnic group to be Takkoi, Hurston then identified that name "not with a population but with a town about forty miles north of Porto-Novo, whose original name is Itakon, and official name, Takon" (Diouf, *Dreams of Africa in Alabama*, 39). In perceiving a linguistic connection between Takkoi and Takon (Itakon), and believing that Burton's account of the destroyed city of Takon was, in fact, an accounting of the destruction of Kossola's hometown, which was destroyed during the same time span, Hurston was convinced that she had discovered a source that corroborated and complemented Kossola's narrative.

The historian Robin Law drew a similar conclusion in *Ouidah: The Social History of a West African Slaving "Port," 1727–1892* (Athens: Ohio University Press, 2004). He writes, "Cudjo was captured in a Dahomian raid on his hometown of 'Togo,' or 'Tarkar,' probably Takon, north of Porto-Novo." In a footnote that references Burton, he adds: "The campaign seems to be identical with that recorded by Burton against 'Attako' (Taccow), near Porto-Novo: *Mission*, I, 256" (138).

Believing that Kossola was of Takkoi ethnicity, convinced that he was from the town of Takon, Hurston was confident in naming Kossola's king "Akia'on." Nevertheless, as Diouf argues, "Cudjo could not have told her it was his king's name" (*Dreams of Africa in Alabama*, 39). Hurston's conclusions in this instance constitute an exception in an otherwise factual manuscript. Diouf writes:

"She may have conflated some of what Cudjo said with some of what she knew as a scholar, but she made a genuine effort at separating the two. With few exceptions, the information provided in *Barracoon* is confirmed by other sources. Witnesses, experts in Yoruba cultures, contemporary newspaper articles, and abundant archival material corroborate the various events in Cudjo's life as described in *Barracoon*." (*Dreams of Africa in Alabama*, 246).]

CHAPTER V

1. Note 1: King "Gelele (bigness), ma nyonzi (with no way of lifting)" (i.e.—too heavy to lift) who ascended the throne in 1858. Gelele succeeded his father, Gezo, at the age of thirty-eight to the exclusion of his older brother, Godo, who was a drunkard.

King Gelele is six feet tall and "looks a king of (Negro) men, without tenderness of heart or weakness of head." Burton's *Mission to Gelele, King of Dahome*, 233.

[Editor's note: Burton, *Mission to Gelele*, 145 n2, 131

n9, 145. (Because there are several editions of the sources Hurston used in her research, I have retained the page numbers original to the manuscript while, within my bracketed editor's notes, I have given the bibliographical data and page numbers of references and citations Hurston made, as I have found them in the particular edition of the works that I have used.)]

2. Note 2: The kings of Dahomey claimed that they never made war upon their weaker neighbors without insult, nor until the war of chastisement was asked for "[f]or three successive years" by the people.—Forbes, *Dahomey and the Dahomans*. [Editor's note: Forbes, *Dahomey and the Dahomans*, 20–21, 15.]

"[S]hould a neighbouring people become rich, it is regarded as sufficient insult to call forth an immediate declaration of war from the court of Dahomey." Forbes, p. 7.

The king of Dahomey, Gelele, said that when his father King Gezo died, he, himself, had "received a message from that chief (Akia'on, King of Takkoi), that all men were now truly joyful, that the sea had dried up, and that the world had seen the bottom of Dahome." Gelele answered by raiding Takkoi and slaying Akia'on, "mounting his skull in a ship (model), meaning that there is still water enough to float the kingdom, and that if the father is dead the son is alive." Burton, *Mission to Gelele, King of Dahome,* pp. 225–26. [Editor's note: Burton, *Mission to Gelele,* 156.]

It is improbable that a weak king would have risked such a provoking message as related by Gelele.

[Editor's note: Although Hurston was mistaken in her conclusion that Takon was the name of Kossola's hometown and Akia'on was the name of its king, the town of Takon did exist and the circumstances surrounding Akia'on's fate as she reports them are accurate.

Having believed Akia'on to be the name of Kossola's king, Hurston thought she had discovered the ultimate cause, *the* "insult," that instigated Glèlè's raid. Whereas she had not questioned Kossola's narration of events of the Dahomian raid on his town, she had questioned his belief that the raid was the result of betrayal by a disgruntled townsman. Perhaps Hurston was also searching for the kind of provocative "reason" that would result in the kind of horrific massacre that Kossola described. Her commentary in note 6 suggests this may have been the case.

The mistake of identifying the circumstances relevant to Takon with those relevant to Bantè would be easy to make given the established modus operandi of the Dahomian king and his warriors.]

3. Note 3: "Industry and agriculture, are not encouraged." The men are wanted for slave hunts.—Forbes,

Dahomey and the Dahomans. [Editor's note: Forbes, *Dahomey and the Dahomans*, 21.]

4. Note 4: "The only other peculiarity in the Court was a row of three large calabashes, ranged on the ground before and a little to the left of royalty. They contain the calvariae of the three chief amongst forty kings, or petty headmen, said to have been destroyed by Gelele," in the first two years of his reign (1858–60); "and they are rarely absent from the royal levees. A European would imagine these relics to be treated with mockery; whereas the contrary is the case. So the King Sinmenkpen (Adahoonzou II) . . . said to Mr. Norris, 'If I should fall into hostile hands, I should wish to be treated with that decency of which I set the example.' The first skull was that of Akia'on, chief of Attako (Taccow) [Takkoi, a Nigerian tribe] near 'Porto Novo,' which was destroyed about three years ago. Beautifully white and polished, it is mounted in a ship or galley of thin brass about a foot long, with two masts, and jibboom, rattlings, anchor, and four portholes on each side, one pair being in the raised quarter deck." The destruction of Takkoi was justified by King Gelele on the ground that King Akia'on had insulted the memory of his father, the late King Gezo.—Burton, *Mission to Gelele, King of Dahome*, pp. 225–26. [Editor's note: Burton, *Mission to Gelele*, 156.]

CHAPTER VI: BARRACOON

1. [Editor's note: Charlotte Osgood Mason funded Hurston's second expedition in the South. Mason would periodically send money to Kossola and would become interested in his general welfare.]

2. Note 5: "The city (of Abomey) is about eight miles in circumference, surrounded by a ditch, about five feet deep, filled with prickly acacia." There are six gates and two grinning skulls are mounted on the gate posts. Inside each gate is "a pile of skulls, human, and of all the beasts of the field, even to the elephant's." The Dahoman standards, each of which was surmounted by a human skull, were much in evidence.—Forbes, *Dahomey and the Dahomans*. [Editor's note: Forbes, *Dahomey and the Dahomans*, 68–69, 73.]

"In the palace at Cannah the legs of the throne rest on the skulls of four conquered princes."—Canot [Editor's note: Hurston may have used another edition of Canot's work in which the statement is a direct quote. In the edition which follows, a similar statement is made about the throne of the king of Dahomey: "Each of its legs rests on the skull of some native king or chief." (Theodore Canot and Brantz Mayer, *Adventures of an African Slaver: Being a True Account of the Life of Captain Theodore Canot, Trader in Gold, Ivory and Slaves on the Coast of Guinea,* ed. Malcolm Cowley (New York: Albert and Charles Boni, [1854] 1928; Whitefish, MT: Kessinger Legacy Reprints, 2012), 260.)]

"The walls of the palace of Dange-lah-cordeh are surmounted, at a distance of twenty feet, with human skulls." Forbes, p. 73. [Editor's note: Forbes, *Dahomey and the Dahomans,* 75.]

Author's note: The author has been informed by natives of Nigeria and Gold coast that it is the custom to carry home the heads of all the people a warrior has killed in battle. He is not allowed to speak of any victory unless he has the heads to show. Mr. Effiom Duke, Calabar district, Nigeria, says that when he was last in Nigeria, less than fifteen years ago, there were skulls lying around so new that the hair was still upon them.

3. (1) There is a festival held in May and June in honor of trade "with music, dancing, singing." (Forbes, *Dahomey and the Dahomans,* p. 16) [Editor's note: This footnote was typed at the bottom of page 55 in the original typescript. See Forbes, *Dahomey and the Dahomans,* 18.]

4. [Editor's note: Kossola's description here of the route he and his compatriots took to the barracoons at Ouidah differs from the route he illustrated for Roche (*Historic Sketches,* 88–89). The routes contradict each other and are otherwise problematic logistically. Diouf suggests two possible explanations: "People coming from different regions took different routes that somehow got conflated when they recounted their march to the sea several decades later; and fading recollections. Although Cudjo had an excellent memory, it is one thing to vividly remember indelible events such as the raid, the march, or the barracoon; it is another to recall the names of towns one had never seen before, especially under such circumstances" (*Dreams of Africa in Alabama*, 49).]

5. Note 6: The native term of derision for the Kroos. They are despised by the other tribes because they are usually the porters for the white men. They are called Many-costs because it is said that many Kroos may be hired for the cost of one decent worker. Some white trader went inland with a number of Kroo porters. While he was doing business with the native king, the porters wandered about the village and arrived at the market place. The girls, as is customary, wore nothing above the waist. The Kroo men amused themselves by pinching the busts of the young women. When the men heard of this desecration they hurried to the headman with the information. He told the white trader to move along with his Kroo porters instantly or they would be killed. The white man replied that the local men could not dispose of his porters because they

were so numerous that the local men were in danger of being chastised themselves. The king replied by asking him, "How many costs?" Meaning, "How much did they cost you?" This was not a question but a sneer, meaning, "They are just as cheap for us to kill as they are for you to hire." Let just one more of your Many-costs pinch titty of our girls and all shall die. The white trader changed his mind and restrained his boys. The story spread and the name stuck to the Kroos.

6. Note 7: Canot, a notorious slave trader, says that the slaves were stripped for cleanliness and health in the middle passage. [Editor's note: Canot and Mayer, *Adventures of an African Slaver*, 108.]

7. [Editor's note: Ibid., 109.]

8. [Editor's note: According to Henry Romeyn's account, in "Little Africa," "One hundred and seventy-five slaves were contracted for. . . . One hundred and sixty-four slaves had been taken on board. Of these but two died on the passage" (15).]

CHAPTER IX: MARRIAGE

1. Note 8: Neither from Kossula nor from the community have I been able to get a clear account of what led up to the killing. One fact is established however: That the community in general feared the Lewis boys.

According to one informant there had been several fights between the Lewis boys and some others, extending over a long period of time. There were numerous grudges to be paid off. The Lewis boys felt as if their backs were against the wall and fought desperately in every encounter.

There was a bloody battle on July 28, 1902, in which one man was shot to death and one seriously wounded with a knife.

Young Cudjo was said to have done both the cutting and the shooting when set upon by some of his enemies. The Negro deputy sheriff is said to have been afraid to attempt an arrest. He tried for three weeks to catch the young man off his guard. Failing in that, he finally approached him concealed in the butcher's wagon and shot young Cudjo to death.

[Editor's note: Note 8 was misnumbered as note 7 in the original manuscript. Hurston's note 7 is a handwritten insertion on the reverse of an early typed manuscript draft, page 28. As Hurston pointed out, the details were not clear. According to Sylviane Diouf and Natalie Robertson, in January 1900, Cudjo Lewis Jr. was convicted of manslaughter in the first degree in the death of Gilbert Thomas, who may have been young Lewis's brother-in-law. Lewis Jr. was condemned to five years in the Jefferson County state penitentiary, but was transferred into the state's convict-lease system. He was pardoned in August 1900.]

CHAPTER XI

1. [Editor's note: Hurston took photographs of Kossola as well as film footage, which can be viewed in Kristy Andersen's PBS American Master's Series production of *Zora Neale Hurston: Jump at de Sun*, 2008.]

APPENDIX

1. [Editor's Note: In "Appendix 3" of *Every Tongue Got to Confess: Negro Folk-tales from the Gulf States*, is a listing of stories Hurston collected from Kossola.]

AFTERWORD

1. Zora Neale Hurston, *Dust Tracks on a Road: An Autobiography* (Urbana: University of Illinois Press, [1942] 1984), 198.

2. Robert E. Hemenway, *Zora Neale Hurston: A Literary Biography* (Urbana: University of Illinois Press, 1980), 95.

3. See Hurston's preface, in the present volume.

4. See Hurston's introduction, in the present volume.

5. Zora Neale Hurston to Carter G. Woodson, July/August 1927, in *Zora Neale Hurston: A Life in*

Letters, ed. Carla Kaplan (New York: Doubleday, 2002), 103; Zora Neale Hurston to Thomas E. Jones, 12 October 1934, in Ibid., 315; Hurston, *Dust Tracks,* 198.

6. A version of this article was published in *The American Mercury* in 1944, and then in a condensed version in *Negro Digest,* also in 1944.

7. Zora Neale Hurston, "Cudjo's Own Story of the Last Slaver," *Journal of Negro History,* October 1927, 648, http://www.jstor.org/stable/2714041.

8. Hemenway, *Zora Neale Hurston,* 96–97, 103 n23.

9. Ibid., 98.

10. Zora Neale Hurston to Thomas E. Jones, 12 October 1934, in Kaplan, *Letters,* 315.

11. Audre Lorde, *Sister Outsider: Essays and Speeches* (New York: The Crossing Press, 1984), 112.

12. Zora Neale Hurston to Thomas E. Jones, 12 October 1934, in Kaplan, *Letters,* 315.

13. Hemenway, *Zora Neale Hurston,* 96.

14. Ibid., 99.

15. Ibid., 98.

16. Valerie Boyd, *Wrapped in Rainbows: The Life of Zora Neale Hurston* (New York: Scribner, 2003), 154.

17. Ibid., 153.

18. Zora Neale Hurston to Langston Hughes, spring-summer 1927, in Kaplan, *Letters,* 99.

19. Boyd, *Wrapped in Rainbows,* 154.

20. Ibid., 153.

21. Hemenway, *Zora Neale Hurston*, 89.

22. Lynda Marion Hill, *Social Rituals and the Verbal Art of Zora Neale Hurston* (Washington, DC: Howard University Press, 1996), 64.

23. Hurston, *Dust Tracks*, 204.

24. Ibid., 200.

25. Ibid.

26. Ibid.

27. Hill, *Social Rituals*, 64.

28. Hemenway, *Zora Neale Hurston*, 100–101.

29. See Hurston's preface in the present volume.

30. See Hurston's introduction in the present volume.

31. Diouf, *Dreams of Africa in Alabama*, 246.

32. Ibid., 246, 3.

33. See Hurston's introduction, in the present volume.

34. Paul E. Lovejoy, *Transformations in Slavery: A History of Slavery in Africa*, 3rd ed. (New York: Cambridge University Press, 2012), 19.

35. Toni Morrison, *Beloved* (New York: Alfred A. Knopf, 1987), i, 275.

36. Diouf, *Dreams of Africa in Alabama*, 66.

37. Philip Curtin, ed., *Africa Remembered: Narratives by West Africans from the Era of the Slave Trade* (Prospect Heights, IL: Waveland Press, [1967] 1997), 9.

38. Morrison, *Beloved*, 274.

39. Hurston, *Dust Tracks*, 204.

40. Marimba Ani (Dona Richards), *Let the Circle Be Unbroken: The Implications of African Spirituality in the Diaspora* (Trenton, NJ: Red Sea Press, 1992), 12.

41. See chapter 8 in the present volume.

42. Henry Romeyn, "Little Africa: The Last Slave Cargo Landed in the United States," in *The Southern Workman* 26.1 (January 1897), 14, http://eds.a.ebscohost.com.ezproxy.lib.usf.edu/eds/ebook.

43. See narrative in present volume.

44. Ibid.

45. Diouf, *Dreams of Africa in Alabama*, 156, 157.

46. Ibid., 2.

47. Ta-Nehisi Coates, *Between the World and Me* (New York: Spiegel and Grau, 2015), 69.

48. James Baldwin, "The White Man's Guilt," in *Baldwin, Collected Essays* (New York: Library of America, [1965] 1998), 723.

Bibliography

Alford, Terry. *Prince among Slaves: The True Story of an African Prince Sold into Slavery in the American South.* New York: Oxford, 1977.

Andersen, Kristy. *Zora Neale Hurston: Jump at de Sun.* Directed by Sam Pollard. PBS, *American Masters*, 2008.

Ani, Marimba (Dona Richards). *Let the Circle Be Unbroken: The Implications of African Spirituality in the Diaspora.* Trenton, NJ: Red Sea Press, 1992.

Boyd, Valerie. *Wrapped in Rainbows: The Life of Zora Neale Hurston.* New York: Scribner, 2003.

Burton, Richard F. *A Mission to Gelele, King of Dahome, Volume 1.* New York: Frederick A. Praeger, [1894] 1966.

Canot, Theodore, and Brantz Mayer. *Adventures of an African Slaver: Being a True Account of the Life of Captain Theodore Canot, Trader in Gold, Ivory and Slaves on the Coast of Guinea*. Edited by Malcolm Cowley. Whitefish, MT: Kessinger Legacy Reprints, [1854] 2012.

Coates, Ta-Nehisi. *Between the World and Me*. New York: Spiegel & Grau, 2015.

Curtin, Philip, ed. *Africa Remembered: Narratives by West Africans from the Era of the Slave Trade*. Prospect Heights, IL: Waveland Press, [1967] 1997.

Diouf, Sylviane A. *Dreams of Africa in Alabama: The Slave Ship "Clotilda" and the Story of the Last Africans Brought to America*. New York: Oxford University Press, 2007.

Equiano, Olaudah. *The Interesting Narrative of the Life of Olaudah Equiano, Written by Himself*. Boston: Bedford Books of St. Martin's Press, [1788] 1995.

Forbes, Frederick Edwyn. *Dahomey and the Dahomans: Being the Journals of Two Missions to the King of Dahomey, and Residence at His Capital, in the Years 1849 and 1850, Volume 1*. Charleston, SC: BiblioBazaar Reproduction Series, [1851] 2008.

Foster, William. "Last Slaver from U.S. to Africa, A.D. 1860." Mobile Public Library, Local History and Genealogy. Mobile, Alabama.

Frost, Diane. *Work and Community among West African Migrant Workers since the Nineteenth Century.* Liverpool: Liverpool University Press, 1999.

Hemenway, Robert. *Zora Neale Hurston, A Literary Biography.* Urbana: University of Illinois Press, 1980.

Hill, Lynda Marion. *Social Rituals and the Verbal Art of Zora Neale Hurston.* Washington, DC: Howard University Press, 1996.

Hurston, Lucy Anne, and the Estate of Zora Neale Hurston. *Speak, So You Can Speak Again: The Life of Zora Neale Hurston.* New York: Doubleday, 2004.

Hurston, Zora Neale. "Barracoon: The Story of the Last 'Black Cargo.'" Typescripts and handwritten draft. 1931. Box 164-186, file #1. Alain Locke Collection, Manuscript Department, Moorland-Spingarn Research Center, Howard University.

———. "Communications," *Journal of Negro History* 12 no. 4 (October 1927). http://www.jstor.org/stable/2714042.

———. "Cudjo's Own Story of the Last Slaver," *Journal of Negro History* 12, no. 4 (October 1927). http://www.jstor.org/stable/2714041.

———. *Dust Tracks on a Road: An Autobiography.* 2nd ed. Urbana: University of Illinois Press, [1942] 1984.

————. "The Last Slave Ship," *American Mercury* 58 (1944), 351–58.

————. "The Last Slave Ship," *Negro Digest* 2 (May 1944), 11–16.

————. *Mules and Men*. Bloomington: Indiana University Press, [1935] 1978.

————. *Tell My Horse: Voodoo and Life in Haiti and Jamaica*. New York: Harper & Row, [1938] 1990.

Jordan, Winthrop. *The White Man's Burden*. New York: Oxford University Press, 1974.

Kaplan, Carla, ed. *Zora Neale Hurston: A Life in Letters*. New York: Doubleday, 2002.

Law, Robin. *Ouidah: The Social History of a West African Slaving "Port," 1727–1892*. Athens: Ohio University Press, 2004.

Lewis, Cudjo (Kossola). Cudjo Lewis to Charlotte Osgood Mason, September 4, 1930. Alain Locke Papers 16499, Moorland-Spingarn Research Center, Howard University.

Locke, Alain LeRoy. *The New Negro: An Interpretation*. New York: A. & C. Boni, 1925.

Lorde, Audre. *Sister Outsider: Essays and Speeches*. New York: Crossing Press Feminist Series, 1984.

Lovejoy, Paul E. *Transformations in Slavery: A History of Slavery in Africa*. 3rd ed. New York: Cambridge University Press, 2012.

Morrison, Toni. *Beloved.* New York: Alfred A. Knopf, 1987.

———. *The Origin of Others.* Cambridge, MA: Harvard University Press, 2017.

Robertson, Natalie S. *The Slave Ship "Clotilda," and the Making of AfricaTown, USA: Spirit of Our Ancestors.* Westport, CT: Praeger, 2008.

Roche, Emma Langdon. *Historic Sketches of the South.* New York: Knickerbocker Press, 1914.

Romeyn, Henry. "'Little Africa': The Last Slave Cargo Landed in the United States," *The Workman* 26. no. 1, Hampton Normal and Agricultural Institute, Hampton, VA, January 1897, 14–17. http://eds.a.ebscohost.com.ezproxy.lib.usf.edu/eds/ebook.

About the Editor

Deborah G. Plant is an independent scholar and writer based in Florida. She is the author of *Every Tub Must Sit on Its Own Bottom: The Philosophy and Politics of Zora Neale Hurston* (1995) and *Zora Neale Hurston: A Biography of the Spirit* (2007), and editor of *The Inside Light: New Critical Essays on Zora Neale Hurston* (2010).

About the Author

Zora Neale Hurston was a novelist, folklorist, and anthropologist. An author of four novels (*Jonah's Gourd Vine*, 1934; *Their Eyes Were Watching God*, 1937; *Moses, Man of the Mountain*, 1939; and *Seraph on the Suwanee*, 1948); two books of folklore (*Mules and Men*, 1935, and *Tell My Horse*, 1938); an autobiography (*Dust Tracks on a Road*, 1942); and more than fifty short stories, essays, and plays. She attended Howard University, Barnard College, and Columbia University, and was a graduate of Barnard College in 1927. She was born on January 7, 1891, in Notasulga, Alabama, and grew up in Eatonville, Florida. She died in Fort Pierce, Florida, in 1960. In 1973, Alice Walker had a headstone placed at her grave site with this epitaph: ZORA NEALE HURSTON: "A GENIUS OF THE SOUTH."